Perspectives in American History

No. 3

THE COTTON KINGDOM IN ALABAMA

THE COTTON KINGDOM IN ALABAMA

By

CHARLES S. DAVIS, PH.D.

ASSISTANT PROFESSOR OF HISTORY
ALABAMA POLYTECHNIC INSTITUTE
AUBURN, ALABAMA

PORCUPINE PRESS
Philadelphia 1974

Library of Congress Cataloging in Publication Data

Davis, Charles Shepard, 1910-
The cotton kingdom in Alabama.

(Perspectives in American history, no. 3)
Originally presented as the author's thesis, Duke University.
Reprint of the 1939 ed. published by the Alabama State Dept. of Archives and History, Montgomery.
Bibliography: p.
1. Cotton growing--Alabama. 2. Alabama--Economic conditions. I. Title.
HD9077.A2D3 1974 338.1'7'35109761 73-16160
ISBN 0-87991-329-0

First edition 1939
(Montgomery, Alabama: Alabama State Department of Archives and History, 1939)

Reprinted 1974 by
PORCUPINE PRESS, INC.
1317 Filbert St.
Philadelphia, Pennsylvania 19107

Manufactured in the United States of America

To

MY WIFE

PREFACE

MY PURPOSE in this study is to describe the principal phases of plantation management, as well as the commercial and financial aspects of cotton planting in Alabama to the outbreak of the Civil War, in the hope that this presentation may be a step toward a more complete understanding of Southern ante-bellum economics.

Hitherto, attempts to evaluate slavery in the South as a system of labor have been directed more toward an internal approach to the institution itself than toward its relationship to the plantation system. The shifting of emphasis in the present study to the economic aspects of cotton production in Alabama has, of necessity, forced me to omit certain moral, political and social phases of slavery which might well be included in a work of wider scope. Only a limited amount of space has been accorded the small farmer and nonslaveholder. Nor has the position of the planter in Southern society been more than mentioned.

This work was begun under the direction of the late Dr. William K. Boyd and carried to completion under the guidance of his successor, Dr. Charles S. Sydnor. To each of these gentlemen I am indebted for inestimable aid and encouragement. I wish to express my sincere appreciation for the suggestions of my colleague, Dr. A. W. Reynolds, and also to Dr. Frank L. Owsley, of Vanderbilt University.

For the loan of private manuscript collections I am indebted to many people in Alabama. Particularly do I wish to thank Mrs. Marie Bankhead Owen, Director of the Alabama State Department of Archives and History, and her staff. Mr. Peter Brannon, State Curator, was especially helpful in locating materials. My thanks are also due to the Duke University Library, Library of Congress, and New York Public Library for their cooperation.

For preparing the maps and charts included in this study I am indebted to Mr. Eugene H. Gray and Mrs. Helen Stowers Womelsdorf, and I wish to express my gratitude to Mr. Arthur H. Tucker, Mr. Wilmer F. Jacob and Professor Henry W. Adams for their cheerful service of reading and editing the text.

Throughout the entire preparation of this study my wife has given me constant aid and encouragement. Particularly do I have her to thank for her painstaking work of editing and typing.

Auburn, Alabama, 1939.

TABLE OF CONTENTS

LIST OF MAPS AND CHARTS

THE COTTON KINGDOM
IN
ALABAMA

CHAPTER I

GEOGRAPHY, SOIL AND CLIMATE

EXTENDING from the shores of the Gulf of Mexico to the Tennessee highlands in the thirty-fifth parallel of latitude, Alabama presents two well-marked geographical divisions which nearly coincide with the northern and southern halves of the State. Each is readily distinguished by climatic differences, topographical features and geological structure, and by the aspect of vegetation corresponding to these conditions. With the exception of the surface irregularities in the northeast, the State may be considered as an undulating plain whose mean elevation is not more than 600 feet. Towards the north and east the surface rises above this elevation, but towards the south and west it sinks considerably below it.[1]

Agriculturally speaking, the State may be divided into six divisions: the Tennessee Valley in the north, the hilly region in the northeast, the Alabama and Tombigbee River basins, the Coosa Valley, the Black Belt in the central part of the State, and the coastal plain which extends to the Gulf of Mexico. Fed by streams which drain the country nearly as far north as the Tennessee Valley, the Alabama and Tombigbee Rivers traverse the central and southwestern portion of the area and empty their waters into Mobile Bay. In ante-bellum days these streams furnished the only commercial highway in the State. They bound the southern section into one cotton-growing com-

[1] Eugene Allen Smith, *Geological Survey of Alabama*, p. 18; Roland Harper, *Economic Botany of Alabama*, Part I, p. 35; Roland Harper, "Preliminary Soil Census of Alabama and West Florida", *Soil Science*, IV, No. 2.

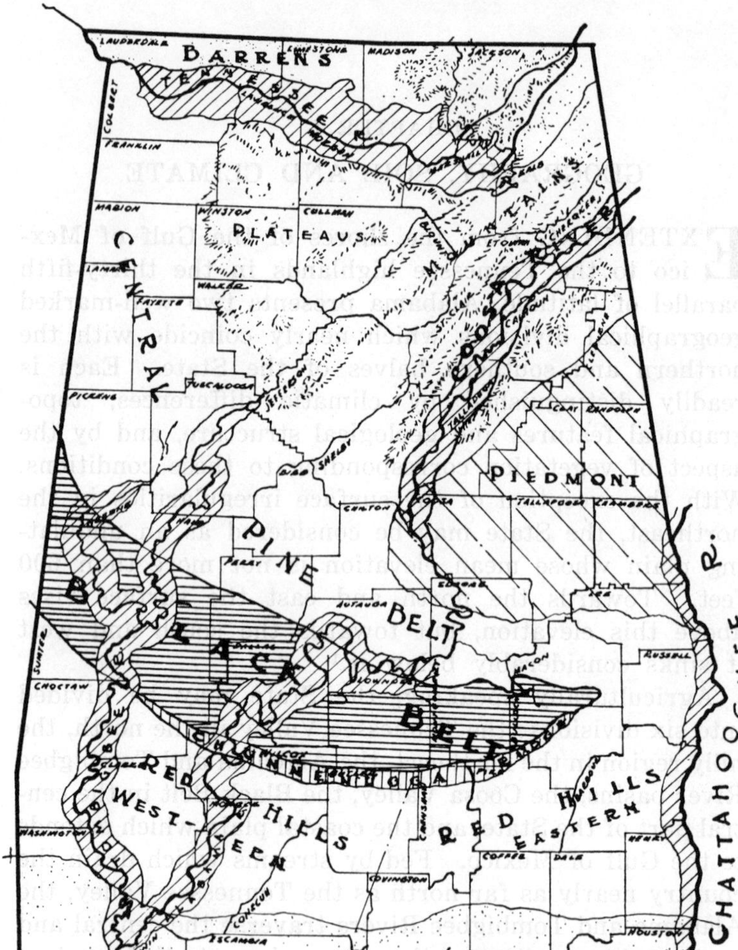

AGRICULTURAL REGIONS OF ALABAMA

munity, and their bottom lands were early sought by the planters of this staple. Making a large bend across the northern end of Alabama, the Tennessee River flows through a wide and fertile valley. During the early days cotton and plantation supplies were transported along the lengthy and tortuous course of this river to its junction with the Ohio. The soil of the valley was fertile, and from the very first attracted planters in large numbers.[2] The Coosa River from Rome, Georgia, down to Greensport, Alabama, was navigable for small craft and subsequently became a connecting link between Northeast Alabama and the Central of Georgia Railroad. Plantations of considerable size gradually came into existence in the rich bottom lands of the valley of the Coosa.[3]

Extremes of temperature in Alabama are comparatively rare because of the moderating effects of the winds from the Gulf of Mexico. In parts of the State most remote from the Gulf, elevation above sea level helps to secure immunity from excessive heat. The mean annual temperature is about 64.5° F., with summer at 79.5° and winter at 50.4°. The springs are early and balmy, the summers are long and even in temperature, the autumns are late and dry, and the winters are so slow to approach and so mild that crops are frequently left in the fields until after Christmas. In the latitude of Montgomery the first frost occurs, as a rule, between the tenth and twenty-fifth of October and the last frost between the fifth and twenty-fifth of April. Since the cotton plant continues to grow and mature its fruit until cut off by frosts, it is obvious that the length of time between frosts becomes an important factor in determining the fitness of a locality for the production of cotton.[4]

---

[2] Eugene Allen Smith, "Cotton Production in Alabama", *Tenth Census*, Cotton, Part 2, VI, 10-12; Thomas Perkins Abernethy, *The Formative Period in Alabama*, pp. 20-25.

[3] *Tenth Census*, Cotton, Part 2, VI, 10.

[4] *Ibid.*, p. 11; Smith, *Geological Survey of Alabama*, p. 26; *The Hill Country of Alabama*, (Anonymous), p. 26.

Annually there is an average of fifty-six inches of rainfall within the State. The rain belt of greatest precipitation narrows in the middle, widens out toward both ends and crosses the State diagonally from southwest to northeast. In the lower part of the belt an area including Mobile and Baldwin and parts of Washington, Clarke, Monroe, Wilcox, Dallas and Lowndes Counties receives a yearly rainfall of sixty-two inches and upward, though sixty-four inches is not uncommon in northeast Mobile County. Records kept in Montgomery show that the average period of wet weather lasts hardly more than five days at a time. During the months of March, April, and May thunderstorms are not unusual and the amount of rain which sometimes falls is great. Snow falls occasionally during the months of January and February. In the lower counties of the State it is extremely rare, but northward there is usually one snowfall of considerable depth during each winter.[5]

Surveyed in more detail, the surface of Alabama is divided into several areas differing in geological formation and fertility of soil. Entering from the northeast and running down towards the center of the State is the southern extremity of the Cumberland plateau. Here the ridges run in long sweeps and give a somewhat mountainous appearance to the region. South of the Tennessee Valley they give way to broken hills which form the north-central portion of the State.[6]

The Tennessee Valley proper averages about fifteen miles in width and stretches entirely across the State from east to west. The region is bordered on both the north and the south by hilly, and in some places almost mountainous, country. The hilly country on the north is known as the barrens and is a part of the great highland rim of Tennessee. South of the barrens lies the valley proper, which has, for the most part, a fertile cal-

[5] *Tenth Census*, Cotton, Part 2, VI, 12.

[6] Abernethy, *The Formative Period in Alabama*, pp. 20-21.

careous soil of a deep red color.[7] The surface is almost level, though occasionally it is broken here and there by slight elevations built up of fragments of chert usually covered with trees. Upon these wooded knolls were frequently situated the dwellings of planters, for in the days when Alabama was at her agricultural zenith the Valley was second only to the Black Belt in agricultural production.[8] The southern border of the Valley is formed by the escarpment of the Warrior coal field, Sand Mountain as it is usually called, rising to an average height of around 600 feet. Generally speaking, the Sand Mountain vicinity was not adapted to raising cotton and therefore lay outside the plantation region.[9]

Skirting the plateau to the east and running parallel to it lies the Coosa Valley, which represents the southern extremity of the great Valley of Virginia. There were stretches of fertile lands here and there which were largely taken up by nonslaveholding farmers who, owing to their isolated position and the difficulty of getting their products to market, were content with small profits. With the coming of rail transportation their main outlet was up the Coosa River to the Georgia railroads at Rome.[10] Grain crops, principally of corn and wheat, seem to have been more profitable than cotton.[11]

Towards the east, between the Blue Ridge and the fall line of the rivers, the Coosa Valley is bordered by a triangular belt of foot-hills averaging about 100 miles in

---

[7] *Tenth Census*, Cotton, Part 2, VI, 28; Harper, *Economic Botany of Alabama*, Part I, pp. 40-41.

[8] Henry McCalley, *North Alabama*, Part I, pp. 2-3.

[9] *Ibid.*, p. 8; Thomas Taylor Jones, "Early History of Madison County", *Alabama Historical Quarterly*, 1930, p. 149; *The Hill Country of Alabama*, p. 32; James Edmonds Saunders, *Early Settlers of Alabama*, pp. 38-39.

[10] *DeBow's Review*, Aug., 1850, p. 220; Abernethy, *The Formative Period in Alabama*, p. 21.

[11] *Matthew P. Blue Papers* (Sketches of Cherokee and Jackson Counties).

width and embracing some 5,000 square miles. Here, as in the Piedmont regions of the older states, the stiff red lands are of unequal quality but with improvement may be made to produce fair average yields. The soils, for the most part, are well supplied with the essential mineral ingredients, except that they are in some instances somewhat deficient in lime. The richest upland soil seems to be in Chambers County. The topography is maturely dissected, and all subject to normal erosion processes, there being no ponds or subterranean drainage such as characterizes the limestone regions. Away from the immediate vicinity of the Blue Ridge, all hills in any neighborhood usually rise to about the same level, and the ridges, as a rule, are somewhat broader and more rounded than the valleys.[12] Until 1836 the Creek Indians retained possession of the territory east of the Coosa River, and consequently this section was not settled as early as other parts of the State.[13]

Beginning at the eastern border a little below the Piedmont and widening considerably in its sweep toward the northwest is what may be appropriately designated as the central pine belt. Its underlying strata, as well as much of the surface deposits, are cretaceous, mostly freshwater deposits, judging from the absence of marine fossils. Here the soil is composed largely of sand and gravel. It is below the average in fertility, and rolling piney woods are characteristic of the region.[14]

Next, toward the south is the Black Belt, prairie or canebrake region, embracing 4,300 square miles in Alabama and extending in a crescent northwestward into Mississippi. The Black Belt is so named because of the dark color of the soil and not because of the overwhelming negro population, as is sometimes thought. Varying

---

[12] Harper, *Economic Botany of Alabama*, Part I, p. 67.

[13] Abernethy, *The Formative Period in Alabama*, p. 21.

[14] Harper, *Economic Botany of Alabama*, Part I, p. 84; R. B. Vance, *Human Geography of the South*, p. 91.

in width from twenty to fifty miles, it embraces fifteen Alabama counties. The region formerly was covered by a thick undergrowth of cane, which was more dense in Marengo County than elsewhere. Hence the term "canebrake", which is often applied to this section.[15]

When settlers first came into the region they were greatly handicapped by the lack of water. The only available sources were ponds and lakes which frequently dried up in the summer, or shallow wells dug in the low damp spots which filled with seepage water and furnished a scant and uncertain supply for drinking.[16]

The soil is composed of rotten limestone with out-croppings of cretaceous lime rock here and there. Holding surface moisture in winter, this soil forms a tenacious mud which renders roads all but impassable, and in the summer it bakes into a hard crust.[17] Where rotten limestone lies at the surface unmixed with drift, it yields on disintegration a gray or greenish gray soil which becomes dark in color when mixed with vegetable matter. The subsoil of the cultivated lands is usually of a lighter color than the topsoil and passes into lime rock at varying depths. From some of the slight elevations the soil has been washed away, leaving bald spots. Here bare rocks partially form the surface of bald prairies, which are not suitable for cotton but produce good crops of corn and oats. The yield of seed cotton on the best prairie land, however, has been estimated at between 800 and 1,200 pounds to the acre.[18]

The soils of the region vary, but all are deficient in

---

15 Edmund Ruffin, *Notes on the Cane-Brake Lands*, p. 4; *Tenth Census*, Part II, Cotton, VI, 46; also see Renwick C. Kennedy, "Black Belt Aristocrats", reprinted from *Social Forces*, Oct., 1934, pp. 1-6.

16 *American Cotton Planter*, New Series, IV, 492-493; Philip Henry Gosse, *Letters from Alabama*, p. 81.

17 For recent description see H. F. Cleland in *Geographical Review*, X, 375-387.

18 See *Tenth Census*, Part II, Cotton, VI, 46-47.

sand and too profusely furnished with carbonate of lime.[19] Edmund Ruffin made a trip through the Alabama Black Belt in the fifties and was convinced that exhaustive methods of agriculture would strip the soil of its fertility. It was generally asserted by many residents that the fertility of the Canebrake would not be depleted by continuous cultivation.[20] The bases assumed for this opinion were the supposedly great depth of the soil and its extreme richness. However, Ruffin believed that no matter how long postponed, exhaustion would come at some time, though he concluded that it was useless to argue this doctrine against those who opposed it.[21]

The Selma chalk, or rotten limestone, differs from most other limestone in Alabama in that it is almost never hard enough to form steep hills or pure enough to be dissolved by percolating waters so as to form lime sinks, caves or subterranean streams which are characteristic of limestone regions. The prevailing topography is gently undulating, in a manner difficult to account for, though probably due almost wholly to normal processes of erosion. In some parts of the region remote from the rivers the land is so level that railroads have built tangents several miles in length.[22]

In spots where LaFayette sandy loam occurs there has been less erosion than elsewhere, though in certain places conical hills have been formed by this process of nature. The rivers which traverse the Black Belt are bordered in most places by steep, bare chalk cliffs of striking appearance. Rivers and creeks, muddy most of the time, fluctuate considerably with the seasons, and many of the small ones practically disappear during the long dry summers.[23]

---

19 Ruffin, *Notes on the Cane-Brake Lands*, p. 6; *Farmers' Register*, II, 182-183.

20 Ruffin, *Notes on the Cane-Brake Lands*, p. 20.

21 *Ibid.*; *The Alabama Beacon*, Dec. 3, 1849.

22 Harper, *Economic Botany of Alabama*, Part I, p. 85.

23 *Ibid.*, pp. 85-86.

The Black Belt is bordered on the south by Chunnennuggee Ridge, the northern half of which shows certain characteristics of black prairie lands while the southern part is made up of brown loam uplands. It runs across the State in a narrow strip, widening somewhat towards the eastern end. A limestone formation underlies this section, but the formation is different from that of the Black Belt in that the stone is of normal hardness. This accounts for the ridges which rise above the prairie in the southern parts of Bullock, Montgomery and Lowndes Counties. The soil is predominately a sandy loam and above the average in fertility. In considering the plantation system and cotton culture, the Black Belt and Chunnennuggee Ridge may be grouped together.[24]

As one proceeds towards the coast, the next significant division of the State is the Southern Red Hills. The eastern part is an upland region with broad divides and narrow swampy valleys, where farming is confined to the smooth uplands. The western part is characterized by narrow ridges and broader valleys.[25] The western division was settled by planters as early as the thirties and has the denser population because of the better soil. In ante-bellum days the contrast in the size of farms of the two sections was quite marked. In 1850 the average size of farms in the eastern part was only 152.5 acres, while 415 acres per farm was the average in the west. Therefore, it is evident that the plantation never developed to any considerable extent on the poorer lands to the east.[26]

The sedimentary strata forming the coastal plain underlie three-fifths of the area of the State. In the lower part, comprising Mobile and Baldwin Counties, the plain

[24] *Ibid.*, p. 91; *Tenth Census*, Part II, Cotton, VI, 48; Abernethy, *The Formative Period in Alabama*, p. 23.

[25] Roland Harper, "Southern Red Hills of Alabama", *South Atlantic Quarterly*, XIX, 204.

[26] *Ibid.*, pp. 204-205; T. H. Ball, *Clarke County—A Glance into the Great Southeast from 1540-1877*, p. 121. Barbour County was perhaps an exception.

rises in gentle swells to an elevation of some 300 feet above the tide-water region. For a distance of from eighty to a hundred miles from the seashore this plain is covered with the mantle of sands and gravel of the LaFayette formation, the oldest of the post-Tertiary strata.[27] This gives rise to soils varying from pure sand on one extreme to sandy loams on the other. The extreme southern part of the region corresponds somewhat to the pine barrens of the South Atlantic states, whereas the northern area is included in the pine belt.[28] In general the soil is poor unless artificially improved by the use of fertilizer. The raising of livestock seems to have been the principal industry of the region before 1860.[29] Some of the extreme southeastern counties, sometimes spoken of as the Wiregrass Area, were dubbed the "cow counties" by newspapers of the fifties.[30]

From the confluence of the Alabama and Tombigbee Rivers to the head of Mobile Bay, which is about thirty miles in a straight line, lies a swampy region covering about 300 square miles. The soils of this delta are alluvia of unknown depth and in most cases well supplied with the essentials of fertility. In upper Mobile and Baldwin Counties bumper crops of cotton and corn were produced in these overflow lands and the plantation system thrived as it did in the Black Belt, though of course on a much smaller scale.[31]

---

[27] Smith, *Geological Survey of Alabama*, p. 19; Thomas M. Owen, *History of Alabama and Dictionary of Alabama Biography*, II, 1254.

[28] Harper, *Economic Botany of Alabama*, Part I, p. 126.

[29] *Matthew P. Blue Papers* (Sketch of Baldwin County).

[30] Lewy Dorman, *Party Politics in Alabama from 1850-1860*, p. 13.

[31] Harper, *Economic Botany of Alabama*, Part I, p. 123; *The Journal of Andrew Ellicott*, p. 201.

CHAPTER II

IMMIGRATION AND EXPANSION

DURING the course of the American Revolution, immigrants began to make their way from the South Atlantic states into the territory which is now Alabama. As early as 1777 the traveler, William Bartram, gave an account of having met a party from Georgia which intended to settle on the Alabama River a few miles above its junction with the Tombigbee.[1] In those early days the most populous settlement in the region, with the exception of Mobile, was on the Tensas River, for here Loyalists driven from North Carolina and Georgia had already begun to collect.[2] In addition to these, the population was made up largely of Indian traders, French half-breeds, and a small group of planters who were interested in raising indigo.[3] Though cotton was by no means unknown, there was little interest manifested in its cultivation until 1795, when Joseph Collins was employed as a surveyor by the Spanish Government at Mobile. On seeing the possibility of large profits from planting he purchased ten slaves from Kentucky, settled them on a tract of rich bottom land near Mobile, and began the cultivation of cotton.[4]

Under the influence of Benjamin Hawkins, the Creeks made considerable progress in agriculture, particularly in the culture of cotton.[5] Hawkins himself records that a

---

1 *Travels of William Bartram*, p. 352.

2 Albert J. Pickett, *History of Alabama and Incidentally of Georgia and Mississippi* (Randolph edition), p. 416.

3 *Ibid.*

4 *Albert J. Pickett Papers.*

5 Lorenzo Dow, *History of the Cosmopolite, or Lorenzo's Journal*, p. 222.

Scotch trader, Robert Grierson, planted upland cotton and employed Indian women to gather it from the fields. However, by 1799 he found it more profitable to manufacture the cotton into coarse cloth than to dispose of it in its raw state.[6]

Even though population slowly increased, in 1800 the Mississippi Territory had only 5,361 whites and 3,489 slaves, and the great majority of these were in the Natchez district.[7] However, by 1805 people in the Mobile region were taking a lively interest in cotton, even though the whole area produced less than 500 bales.[8]

The first cotton gin in the eastern portion of the Mississippi Territory seems to have been constructed by a Pennsylvania Jew, Abram Mordecai. In 1804 he erected a ginhouse at the junction of the Coosa and Tallapoosa Rivers, which was an advantageous location for Indians bringing raw cotton down the rivers by canoe. Some of the first bales are said to have gone to the nephew of Governor Defort at New Orleans and brought a price of thirty-three cents a pound.[9] Three years later, cotton receipts obtained from the owner of a gin were made legal tender and passed as domestic bills of exchange.[10]

Nevertheless, the Tombigbee planters suffered greatly as a result of greater transportation expenses and double tariff charges they were forced to bear. Goods entering the region had to pass through the Spanish customs house at Mobile and the American port of entry at Fort Stoddert. This meant that a duty of from 45% to 47% was collected on domestic articles necessary for family comfort, while at the same time people in the Natchez district

---

[6] Benjamin Hawkins, "A Sketch of the Creek Country", *Collections of the Georgia Historical Society*, III, 44.

[7] Adam Seybert, *Statistical Annals*, p. 21.

[8] G. B. Avant, *History of Washington County Alabama to 1860*, p. 22 (M. A. Thesis, U. of Ala.); James L. Watkins, *King Cotton—A Historical and Statistical Review*, p. 137.

[9] *Albert J. Pickett Papers.*

[10] Pickett, *History of Alabama*, p. 503.

were entirely free from such exactions. There flour from Kentucky sold for four dollars a barrel, while in the Mobile district the planter paid sixteen.[11]

Between the years 1805 and 1809 slave owners from North Carolina, Georgia and Virginia began to move into the Tennessee Valley. In December, 1808, Governor Williams issued a proclamation forming part of the region into Madison County with Huntsville as the county seat.[12] One group of immigrants, largely from Virginia and Pennsylvania, had previously been engaged in the culture of tobacco in Georgia, but the profits in upland cotton soon made it desirable for them to abandon their former occupation and move into the Tennessee Valley to take up cotton lands.[13] The crop of 1809 amounted to approximately 1,500 bales for the whole Alabama region, and probably much less than half came from the vicinity of Mobile.[14]

The whole Mississippi Territory continued to develop. In the words of the historian Pickett

> The forests began to be extensively felled; houses were reared as if by magic; the preacher was zealous in the discharge of his divine mission; the 'schoolmaster was abroad'; the medical and legal professions flourished; the merchant drove a good business; the mechanics received constant employment and high wages — while the farmer worked for them all and received his due reward.[15]

There is no doubt, however, that the settlements on the Mississippi River had reached a state of development far in advance of their eastern neighbors. The census of 1810 showed a combined white and slave population of over 40,000 for the whole Territory.[16]

---

11 *Ibid.*

12 *Ibid.*, p. 505.

13 Thomas Perkins Abernethy, *The Formative Period in Alabama*, p. 24.

14 Watkins, *King Cotton*, p. 137.

15 Pickett, *History of Alabama*, p. 505.

16 Seybert, *Statistical Annals*, p. 22.

Cotton was early introduced into the Mississippi Territory and the spread of its culture over the Southwest appeared inevitable. This movement was somewhat delayed by the War of 1812 and the fact that certain areas were still occupied by the Indians. However, after the war a scarcity of cotton resulting in extremely high prices caused a renewed emphasis on the production of this staple. When normal relations were restored by the treaty of peace, England again obtained her supply of American cotton, and in 1815 the price rose to almost thirty cents a pound.[17]

The war also brought another economic crisis, which again hastened the developments already in progress. The Atlantic seaboard began to be depopulated because of the low price of tobacco due to the temporary loss of the British market, trade restrictions, and competition from the West Indies. In the meanwhile, Jackson had subdued the Creeks and thus thrown open a new region of Alabama for settlement.[18] Cotton was excellently adapted for production by the plantation system, and as the effects of soil exhaustion began to become evident in the Upper South, individual planters turned to the Southwest in search of new lands. The small producers, being at a greater disadvantage, were usually the first to make the change. The tide of small farmers, advancing toward the frontier in search of new opportunity, was followed by the planters who sought new openings where capital might be more advantageously invested than in the older districts.[19]

In certain districts where the soil was less fertile or facilities for transportation were not available, the small farmer was allowed to go his own way unhampered.

---

[17] Abernethy, *The Formative Period in Alabama*, p. 25.

[18] Ulrich B. Phillips, *Plantation and Frontier Documents*, I, 87; R. S. Cotterill, *The Old South*, p. 120.

[19] Ulrich B. Phillips, "Growth of Southern Black Belts", *American Historical Review*, XI, 2 (Reprint).

However, in the more desirable locations a conflict arose between the planter and the frontiersman. In some instances the pioneer moved on, and in others, by energy and thrift, he saved enough money to buy slaves and thus became a planter.[20] The planter continued to encroach upon the territory occupied by the small farmer, and Professor Phillips stated that "the very nature of the plantation system caused this phenomenon". One suitable area after another was occupied by much the same process. Planters continued to compete among themselves and wore out such soil as they came to acquire, then moved on in search of new regions to exploit.[21]

In the years between 1820 and 1830 no section underwent any more far-reaching change than did the states of the Atlantic seaboard. The invention of the gin had now made profitable the cultivation of short-staple cotton, and the factory system of Great Britain, made possible by the inventions of Arkwright, Hargreaves and Cartwright, furnished the means of manufacturing cotton cloth beyond the world's supply of raw material. Under the stimulus of this demand for cotton, year by year the plantation area was extended westward. In the twenties many of the southern counties of Virginia were attempting cotton cultivation, interior counties of North Carolina were combining the culture of cotton with the older industries, and in South Carolina the area of cotton and slavery had extended up the rivers well beyond the middle of the state. Even in Georgia, the cotton planters so long restrained by the Indian line broke through the barriers and spread all over the newly-ceded lands.[22]

The Georgia and South Carolina Piedmont regions had

---

[20] *Ibid.*, pp. 2-3; "The Autobiography of Gideon Lincecum", *Mississippi Historical Society Publications*, VIII, 4.

[21] *Ibid.*, VIII, 3. The Panic of 1819 was also a factor in the immigration movement to the Southwest.

[22] Frederick J. Turner, "The South 1820-1830", *American Historical Review*, XI, 559.

two distinct disadvantages from the point of view of the cotton planter. The soil was not as well adapted for the production of cotton as were the Black Belt and river bottoms of Alabama, and the whole region lacked transportation facilities as it was cut off from the tidewater by the pine barrens through which passed few navigable rivers. Thus cotton declined in the old states of the South in proportion to its expansion into the Southwest. In 1800 three-fourths of the nation's output came from Virginia and the Carolinas, while sixty years later two-thirds was produced in Georgia, Alabama, Mississippi and Louisiana. Population began to flow from the older states into the new country so that the drain was keenly felt in certain deserted communities.[23] James Graham of Hillsborough, North Carolina, stated in a letter to Thomas Ruffin in 1817:

> I have it in contemplation to explore the Mississippi State and Alabama Territory next fall or winter so as to satisfy myself respecting the advantages and inducements which that country is said to afford. The *Alabama Fever* rages here with great violence and has carried off vast numbers of our citizens. I am apprehensive, if it continues to spread as it has done, it will almost depopulate the country . . . . Some of our oldest and most wealthy men are offering their possessions for sale and are desirous of removing to this new country.[24]

The Upper South was partially compensated for its loss in the shift of population and capital by the demand for surplus slaves in the cotton fields of the Southwest.[25]

---

[23] Abernethy, *The Formative Period in Alabama*, p. 25; Walter Fleming, "The Slave Labor System in the Ante-Bellum South", *The South in the Building of a Nation*, V, 113; Charles W. Ramsdell, "The Natural Limits of Slavery Expansion", *Mississippi Valley Historical Review*, XVI, 153.

[24] J. G. de R. Hamilton, ed., *The Papers of Thomas Ruffin*, I, 198.

[25] Frederick J. Turner, *American Historical Review*, XI, 561; Ernest Bogart, *The Economic History of the United States*, pp. 135-137.

Streams of immigrants poured into the Alabama region from several directions. The rich valley of the Tennessee River was settled largely from Tennessee, and indirectly through Tennessee from the older states. In the central portion, along the rivers, settlers came chiefly from Virginia and the Carolinas, but in many instances stopped over in Georgia a few years on their way westward.[26] Charles and James Tait of Wilcox County came from Virginia to Elbert County, Georgia, sometime between the close of the Revolution and 1810 and then moved into Alabama in 1818.[27] When Charles Tait moved to Alabama he had twenty-five working hands, and the next year he and his son, James, put their slaves together and used the proceeds from the crop to buy more land. "In 1820 we cleared the upland afterwards known as the Old Place and made 80 bales of cotton, about double what we had made in Georgia with the same force, except that our young hands were two or three years older."[28]

The eastern portion of the State and the Mobile district were settled by people from many different states, though largely from Georgia. One colony, consisting of French exiles who had followed the fortunes of Napoleon until his downfall, founded the town of Demopolis on the Tombigbee.[29] At the same time that population was crowding into the country north of the Tennessee River into that portion of the Chickasaw and Cherokee territory afterwards organized into Jackson, Limestone and Cherokee Counties, the country to the south of Madison County was likewise receiving its advance of pioneer settlements in

---

26 George Petrie, "Alabama 1819-1865", *The South in the Building of a Nation,* II, 271.

27 *James A. Tait Plantation Memorandum Book.*

28 *Ibid.*

29 Gaius Whitfield, Jr., "The French Grant in Alabama, A History of the Founding of Demopolis", *Transactions of the Alabama Historical Society,* IV, 331-355; Pickett, *History of Alabama,* pp. 623-634.

all that portion of the Tennessee Valley which now comprises the counties of Franklin, Lawrence and Morgan. Nor was this the limit of immigration, as hundreds were advancing down the Tombigbee to the settlements on the lower portion of the river near Washington County and others were advancing westward up the headwaters of the Tombigbee, taking up the virgin lands still in the occupancy of the Chickasaws.[30]

This tide of immigration toward the headwaters of the Tombigbee and Black Warrior Rivers was encroaching upon the contiguous territories of the Choctaws, Chickasaws and Cherokee nations, who were in friendly alliance with the United States.[31] To facilitate the advancement of these settlements, chiefly north and east of the Creek nation, the Federal Government took immediate measures to obtain a formal relinquishment of the claims of the three Indian nations. For this purpose commissioners were appointed on the part of the United States, who, during the autumn of the year 1816, concluded three separate treaties for the cession of all the territory from the headwaters of the Coosa westward to the Tombigbee at Cotton-gin Port, and to a line running thence direct to the mouth of Caney Creek on the Tennessee River. These were the last treaties for the relinquishment of Indian lands within the Mississippi Territory previous to its division into two separate governments.[32]

Though the treaty of Fort Jackson and the Indian cessions of 1816 opened up a vast area for settlement, agricultural expansion in Alabama remained restricted until the removal of the Indians west of the Mississippi became complete in 1836. George S. Gaines was appointed as agent of the United States Government in 1831 to superintend the removal of the Choctaws, who, along with the

---

[30] John W. Monette, *History of the Discovery and Settlement of the Valley of the Mississippi*, II, 433.

[31] *Ibid.*, p. 436.

[32] *Ibid.*

Chickasaws, departed peaceably.[33] The Creeks and Cherokees steadfastly refused to cede and opposed with all the means in their power the final treaties of cession. However, by the treaties of Cusseta in 1832 and New Echota in 1835 both the Creeks and the Cherokees were induced to cede to the United States Government all of their remaining lands east of the Mississippi River.[34]

On March 1, 1817, Congress declared that the Mississippi Territory should be divided by a line commencing at the mouth of Bear Creek on the Tennessee, thence to the northwest corner of Washington County, and thence south to the Mississippi Sound. A convention composed of forty-four delegates assembled at the town of Washington, near Natchez, to adopt a constitution, but none of the counties now lying within Alabama was represented at this convention.[35]

The settlement of Alabama during the decade after 1817 was indeed rapid. The flood gates of immigration were hoisted and the human tide of settlers poured in from Virginia, the Carolinas, Tennessee and Georgia. During the most favorable seasons for journeying, the ferries along the Chattahoochee were crowded with immigrant trains. Not infrequently a family was held up for several days because of the great numbers who had to be carried across.[36] The rush was so great in the spring of 1817 as to threaten many with starvation unless supplies were sent in from other parts. The price of provisions skyrocketed. At Fort Claiborne flour was selling at twenty dollars a barrel and corn at five dollars a bushel.[37]

---

33 *Mobile Commercial Register*, Sept. 10, Nov. 12, Dec. 24, 1831.

34 *Ibid.*, Sept. 27, Oct. 27, 1829; Grant Foreman, *Indian Removal*, pp. 114-115.

35 Monette, *History of the Discovery and Settlement of the Valley of the Mississippi*, II, 437; Pickett, *History of Alabama*, p. 614.

36 Albert J. Pickett, *The Southern Cultivator*, V, 10; B. F. Riley, *History of Conecuh County*, p. 73.

37 William Garrott Brown, *A History of Alabama*, p. 130.

By 1820 the State had grown into a commonwealth of 127,901 inhabitants, about 85,000 of whom were whites and 42,000 were slaves.[38] The Great Bend of the Tennessee River was the most densely populated portion of the State. Madison County with 17,481 and Limestone with nearly 10,000 were the largest counties from the standpoint of population.[39] The Black Belt counties contained fewer people than those in the Tennessee Valley and in every case slaves made up at least half of the population.[40] With the increase in the number of inhabitants came a corresponding increase in the sale of public lands and the export of cotton. From 1817 to 1822, ten million dollars worth of public lands were purchased in Alabama, and in 1821 cotton exports were estimated at three millions.[41]

In coming from the older states into Alabama the immigrants followed several rather well-defined routes. Two roads through the Mississippi Territory were continuations of established routes of travel. The one from Nashville to Natchez crossed the Tennessee River at Muscle Shoals and was known as the "Natchez Trace". This was a continuation of the "Kentucky Trace" which ran from Lexington southward to Nashville. Northward from Athens, Georgia, a highway passed through Greenville, Charlotte, Fredericksburg, then on to Washington.[42] The main route south from Athens to New Orleans followed the general direction of the Alabama River, passing through St. Stephens in Washington County and then turning west to the Mississippi River. This came to be known as the "Federal Road". Though appropriations

---

38 *Fourth Census of the United States*, pp. 22-23.

39 *Ibid.*

40 *Ibid.*

41 *Niles Register*, XXI, 381; *Mobile Commercial Register*, Feb. 25, 1822; *Alabama Republican*, Nov. 24, 1820; Feb. 8, 1822.

42 Abernethy, *The Formative Period in Alabama*, p. 29; Frederic L. Paxon, *History of the American Frontier*, p. 209.

for the road were made by Congress in 1806, it was not opened for travel until 1811.[43] Along this road came such families as that of Israel Pickens who settled in Greene County, the Taits and Burfords of Wilcox, and many others who settled the central and southern portions of the State.[44] In January, 1817, Israel Pickens, writing to General William Lenoir in North Carolina, said, "My family have reached our destination in good health without any accident or misfortune. The journey lasted six weeks and the roads were bad beyond conception. . . ."[45] Hodgson records that in 1820, along this same road, he met several gangs of slaves accompanied by their masters, moving into Alabama. Towards evening of the same day he met six wagons conveying ninety slaves belonging to a Georgia planter who had taken up lands in Alabama on the Cahaba River.[46]

Diverging from the upper Piedmont route just beyond the Georgia line was the fall line road which ran from Milledgeville to Augusta, Columbia, Raleigh and Richmond and connected with the upper road just before reaching Washington. Planters from the tidewater of Virginia and North Carolina and the low country of South Carolina found this a convenient route by which to reach the Southwest.[47]

The Great Valley of Virginia presented still a fourth route which was an important means of reaching Alabama. The road, branching off the Pittsburgh-Philadelphia highway, passed southwestward to Knoxville. From Knoxville the original highway turned west to Nashville,

---

[43] Ulrich B. Phillips, *History of Transportation in the Eastern Cotton Belt to 1860*, p. 69; T. H. Ball, *Clarke County — A Glance into the Great Southeast from 1540-1877*, p. 134.

[44] See the *Pickens Papers*, *Burford Papers*, *Tait Papers* and *Alabama Historical Quarterly*, Spring issue, 1930.

[45] *Israel Pickens Papers.*

[46] Adam Hodgson, *Letters from North America*, I, 113.

[47] Phillips, *Transportation in the Eastern Cotton Belt*, pp. 68-69.

but with the rapid development of Madison County, a spur which was soon to become an important route into the Tennessee Valley extended down to Huntsville.[48]

Though it is impossible to speak with absolute authority on the distribution of immigrants in Alabama, there are sections where influences of certain of the older states are more pronounced than in others. For instance, in the Tennessee Valley, somewhat isolated from the rest of the State yet easily accessible to men across the line, Tennesseans came to predominate. Some took up lands in the Valley and others passed on southwest through Jones Valley to settle in the out-of-way places where frontier life existed for a long time.[49] Finally the Tennesseans came to predominate in Blount County and made up a considerable part of the population in Jefferson.[50]

In most cases each community had its own local characteristics, modeled in a measure after the state from which the majority of the settlers came. In the Tombigbee and Alabama basins, as well as in the Black Belt, a majority of the settlers seem to have come from Virginia and the Carolinas, though Georgia and Tennessee were by no means without their representatives. Speaking of the Virginian, Baldwin writes:

> He never gets acclimated elsewhere; he never loses citizenship to the old home. The right of expatriation is a pure abstraction to him. He may breathe in Alabama, but he lives in Virginia. His treasure is there, and his heart also. If he looks at the Delta of the Mississippi, it reminds him of the James River's 'low grounds'; if he sees the vast

48 *Ibid.*; Abernethy, *The Formative Period in Alabama*, p. 29; Alderman and Gordon, *J. L. M. Curry, A Biography*, p. 38; John Robert Kennamer, *History of Jackson County*, p. 11.

49 *DeBow's Review*, Jan., 1855, p. 23.

50 George Powell, "History of Blount County", *Alabama Historical Society Publications*, 1855, p. 37; D. B. Warden, *Statistical, Political and Historical Account of the United States of America*, III, 34; R. A. McClellan, *Early History of Limestone County*, p. 15.

prairies of Texas, it is a memorial of the meadows of the Valley.[51]

During the first years of settlement North Carolinians apparently showed a preference for Greene County,[52] while settlers from South Carolina seem to have preferred Pickens.[53] However, according to Snedecor's *Directory,* compiled in 1856, Greene County showed 438 native South Carolinians and 348 North Carolinians, with a scattering from Georgia, Tennessee and Kentucky.[54] In Marengo County, which is just south of Greene, settlers from both the Carolinas seem to have been about evenly divided. In 1860 there appear to have been 294 from North Carolina and 213 from South Carolina, while Virginia and Georgia were represented with approximately 150 each.[55]

Though most of the immigrants to this region of the State apparently came overland along one or the other rather well-defined routes, there is evidence to show that some came by sea from the Atlantic ports to Mobile and thence up the rivers. Joseph Stickney, a French grantee who settled a few miles south of Greensboro, returned to North Carolina in 1815, chartered a schooner, and brought his family and slaves by water to Mobile and thence up the river.[56]

Eastern and Southern Alabama, with the exception of the Mobile area, were populated chiefly from the State of Georgia.[57] In some instances settlers originally came

---

[51] Joseph G. Baldwin, *Flush Times in Alabama and Mississippi,* p. 73.

[52] Henry Watson, Jr., to J. A. Reed, Feb. 12, 1834, *Watson MSS.*; W. E. Yerby, *History of Greensboro,* p. 2.

[53] Nelson F. Smith, *History of Pickens County,* pp. 36-37.

[54] G. V. Snedecor, *Directory of Greene County,* 1856.

[55] W. C. Tharin, *Directory of Marengo County,* 1860, p. 44; J. W. DuBose, *Chronicles of the Canebrake.*

[56] *Ibid.*

[57] Ball, *Clarke County,* p. 131; T. H. Jack, *Sectionalism and Party Politics in Alabama,* p. 10; Brown, *A History of Alabama,* p. 130.

from other states, merely settling temporarily in Georgia before moving to Alabama. Such was the case of Charles and James Tait of Wilcox, and Henry Lucas of Montgomery County.[58] During the winter of 1817 a large number of Georgia immigrants coming by way of the Federal Road settled in Clarke, Butler, Monroe and Conecuh Counties.[59] Among this group were John Eads and family on their way to Louisiana, but owing to the scarcity of provisions they decided to stop on the bank of the Alabama to raise a crop of corn and rest their teams. The river bottoms were so productive that Eads remained permanently in the neighborhood of Carney's Ferry.[60]

The region around Mobile was the most cosmopolitan in the entire State. Here marked influences of both the French and Spanish occupations were clearly discernible, particularly in the half-breed population which was usually called Creole. The merchants and traders were in most instances Northerners, many coming from New England,[61] while the common people in the Tombigbee settlements had come principally from Georgia.

The tide of immigration pouring into Alabama between 1815 and 1820 gave rise to inflated land prices and an orgy of speculation only to be cut short by the Panic of 1837. The financial situation of the country was such as to promote the rapid rise of land prices, since during the War of 1812 there had been a general suspension of specie payments, and the currency of the country had fallen into great disorder. To make matters worse, Alabama lands, in many instances, could be purchased with Yazoo script.[62] Many bank notes which circulated were

---

[58] W. G. Robinson, *Recollections of the Early Settlers of Montgomery County*, p. 42.

[59] John Buckner Little, *History of Butler County*, pp. 23-25.

[60] *Extract from the Journal of John Eads.*

[61] A. B. Meek, *Romantic Passages in Southwestern History*, p. 241.

[62] Thomas Jones Taylor, "Early History of Madison County", *Alabama Historical Quarterly*, 1930, p. 101.

of uncertain value and usually depreciated in proportion to the distance from the place at which they were issued. In North Alabama, notes from Tennessee banks circulated freely, while in the southern part of the State Georgia bank notes were found in abundance. Currencies between north and south were not interchangeable, because the notes circulated at par only in the region where the bank was located. Thus commercial transactions between the two sections were extremely limited.[63]

Even before the division of the Mississippi Territory, marked evidences of sectionalism between the northern and southern parts of the region were apparent. Because they lacked transportation facilities to a market within the State, North Alabamians were forced to launch their cotton upon the dangerous and expensive journey down the Tennessee River, and after a hazardous trip down the Mississippi sell in a market 1,500 miles from home. This commercial estrangement from South Alabama carried with it social and political consequences. The two sections of the State were one in name only and did not have that identity of interest and feeling which is essential to prosperity and political harmony.[64] As early as 1819 this state of affairs was brought to the attention of the people of Alabama by Governor Bibb, who suggested that the extremes of the State might be connected by means of a turnpike or by a canal running from the Tennessee River to the headwaters of the Alabama.[65]

From 1817 to 1819 unlimited credit and an abundance of paper money gave rise to an unprecedented land boom in Alabama. Having been surveyed, the land was advertised for sale at public auction in various parts of the State. Land offices had been opened at St. Stephens in

---

[63] C. H. Alexander, *The Bank of the State of Alabama and its Branches* (M. A. Thesis, U. of Ala., 1925), p. 7; Riley, *History of Conecuh County*, p. 25.

[64] *DeBow's Review*, Feb., 1850, p. 179.

[65] *Ibid.*

1803 and at Huntsville in 1807 and were destined to be the centers of the great boom of 1819 and 1820.[66] Tracts were usually sold to the highest bidder, and those remaining unsold might be entered privately at a minimum price of two dollars per acre, but in either case one-fourth of the purchase price had to be paid at the time of the sale, and the remainder in four annual installments.[67] The number of sales in 1810 amounted to fifty-three. 1,602 acres were sold at Huntsville and about the same number at St. Stephens, the average price being $2.20 per acre.[68] Early deed records of Madison County indicate clearly that during the next decade land prices rose until they were more than double those of 1810.[69]

After the Indian cessions of 1816, speculators and land sharks began to invade the Tennessee Valley in great numbers.[70] Fictitious and fraudulent methods were used by the speculators to keep out *bona fide* settlers. When lands were offered for sale, speculators ran the prices up so high that those who were purchasing tracts for cultivation were forced to pay far more than they could afford. Later, when settlers were unable to make their payments, these lands were thrown on the hands of the Federal Government and were bought by the speculators for a small initial payment at the land office.[71]

Prominent among the speculators of North Alabama was General John Coffee, who previously had been appointed principal surveyor of the region. He had the

---

66 *American State Papers, Public Lands*, VII, 543.

67 George L. Sioussat, *The Public Land System of the State and Federal Governments in the South*, pp. 70-71.

68 *Correspondence of the Secretary of State Transmitting a Statement of the Sales of Public Lands in Alabama*, 1809-1818; *American State Papers, Public Lands*, V, 385.

69 See *Deed Records*, Madison County, Book A.

70 Payson J. Treat, *The National Land System*, p. 176; James C. Ballagh, *Annual Report of the American Historical Association, 1898*, p. 258.

71 *Ibid.*

advantage of knowing beforehand the choice tracts which were being thrown open for sale, and profited greatly from this knowledge. Employees of the land offices were bonded and sworn to the oath that "in consideration of the advantages to us as above stated, we agree and oblige ourselves severally . . . that we will not impart information to others to enable them to purchase lands gratuitously or without a valuable consideration".[72] Nevertheless, Coffee organized the Cyprus Land Company and succeeded in getting the site he desired at the foot of Muscle Shoals, where he laid out the town of Florence. Andrew Jackson owned stock in the enterprise and kept in close touch with Coffee.[73] The surveyor-general apparently was also involved in the Muscle Shoals and York Bluff land companies,[74] and in March of 1819 he purchased in Limestone County, in a joint scheme with John Read, 188 acres at fifty dollars an acre for purposes of speculation.[75]

In most instances, purchasers signed a series of notes to cover the payments to the land companies. If the notes fell due and the purchasers were unable to pay, they were sold out. This seems to have occurred many times in Limestone and Madison Counties between 1819 and 1825, when all properties were sold to satisfy payment — crops in the fields, household furnishings and negroes, if there were any.[76] Such a fate befell William E. Gilbert of Washington, Georgia, who made an advance payment for a tract of land in Madison County in 1819. Unable to honor his notes when they fell due, his creditors auctioned off his household furnishings for $3,100 and took a mort-

---

[72] *John Coffee Papers.*

[73] *Ibid.*; Thomas Perkins Abernethy, *From Frontier to Plantation in Tennessee*, pp. 271-272.

[74] *John Coffee Papers.*

[75] *Deed Records of Limestone County*, 1819-1825.

[76] *Ibid.*

gage on his six negroes, who were later sold in the same manner.[77]

The sales from 1817 to 1819 were the largest the Tennessee Valley witnessed during the period of settlement and expansion. In February, 1818, all lands lying west of Madison County on both sides of the Tennessee River were placed on the market. Competition for the good lands, which brought enormous prices ranging from twenty to fifty dollars an acre, was keen. River bottoms sold for more, in some cases even bringing as much as $100 per acre.[78] In all, 1,003,423 acres were disposed of at an average price of $6.95 per acre.[79]

Never again was such a vast amount of good land put on the market in Alabama, but during the same years similar booms were taking place at Cahaba and St. Stephens. The Federal land office was moved from Milledgeville to Cahaba in 1815,[80] and owing to the fact that the General Assembly had authorized the removal of the permanent seat of government to that place, town lots rose rapidly in price. In a transaction of May 1, 1819, Nicholas Cochran sold to John Punaldin a town lot for $5,025,[81] and in October of the same year Governor Bibb stated that $123,856 had been taken up for 182 town lots.[82]

The Receivers Balance Account Book indicates that many prominent people took part in the speculations at Cahaba. In 1818, Governor William Wyatt Bibb took up 600 acres at six dollars per acre, and paid $855 in certificates of Mississippi stock. Bolling Hall bought 400 acres

---

[77] *Deed Records*, Madison County, Book E.

[78] Taylor, "Early History of Madison County", *Alabama Historical Quarterly*, 1930, p. 490; *American Quarterly Register*, X, 2077.

[79] *American State Papers, Public Lands*, V, 383.

[80] *Ibid.*, VII, 543.

[81] *Deed Records*, Dallas County, Book A.

[82] Extract from the *Nashville Whig*, Feb. 27, 1819, *John Coffee Papers*.

and made his down payment in Mississippi certificates. Other persons who bought land for speculation were William Rufus King, Henry Lucas, Alexander Pope and numerous persons from Georgia and North Carolina.[83] Prices ranged from two to thirty dollars per acre,[84] and during the course of a year transactions involving over a million dollars passed through the land office.[85]

On a smaller scale, sales progressed at St. Stephens, where people from the adjoining counties came to buy river bottom lands. The average price was, as a rule, lower than at Huntsville and Cahaba, and the average size of the purchase was about eighty acres.[86] Israel Pickens wrote General Lenoir in North Carolina, "I have bought a tract of land on the Tombigbee fifteen miles below St. Stephens for $4.00 per acre."[87] However, the speculative mania never reached the height at St. Stephens that it did in the northern and central portions of the State.[88]

When the tide of speculation had receded, the people of Alabama were left with a debt of over $11,000,000 for land purchased on credit.[89] The price of cotton had fallen off sharply and the country was in the throes of a commercial depression. The prospect of paying for land bought at abnormal prices was now hopeless. Appeals were sent to Congress, and in 1820 a bill was passed which abolished the credit system and reduced the size of the tracts offered for sale to eighty acres.[90] The law allowed purchasers to relinquish lands and apply the payments on the part they retained, the other payments to

---

83 *Receivers Balance Account Book*, Cahaba, 1819.

84 *Ibid.*; *Register of Certificates*, Cahaba, 1820-1835.

85 *American State Papers, Public Lands*, V, 383.

86 *Register of Certificates*, St. Stephens, 1820-1845.

87 *Israel Pickens Papers*, Dec. 27, 1818.

88 *American State Papers, Public Lands*, V, 383.

89 *Ibid.*, V, 378-380.

90 *Alabama Republican*, Nov. 3, 1820; Benjamin H. Hibbard, *A History of Public Land Policies*, p. 98.

be made at a discount of 37½% in installments covering from six to eight years, without interest.[91]

The trustees of the Cyprus Land Company agreed to extend the benefit of the relief act to those who had purchased land provided they applied to General Coffee or James Jackson on or before December 30, 1821. The old bonds of the land company were to be replaced by a new issue payable in four equal, annual installments with a deduction of 20% and a reduction of all interest which might then have accrued. "Those who do not avail themselves of the benefit of this law or otherwise settle with the Trustees will be sued indiscriminately."[92]

In 1826 the people of Alabama petitioned Congress for a law which would divide unsold land into classes with each class given a price according to its grade.[93] This was to prohibit speculators from bidding against the settler who was trying to buy the land on which he had already made improvements. Nothing, however, came of the plan, but instead a law was passed which permitted those who had relinquished lands to repurchase them at a reduction of 37½% of the original price.[94] This did not relieve the situation, and the fear of being sold out still haunted the small land owner. At Cahaba farmers entered written agreements to prevent speculators from bidding against foreclosed land. On several occasions open clashes occurred, during which several speculators were driven from the town and others were threatened.[95]

Because private banking interests had done much to foster the spirit of speculation in the State by large issues of notes improperly secured, in 1820 the citizens of Madi-

---

91 *American State Papers, Public Lands*, VI, 636; *Alabama Republican*, Nov. 10, 1820; William R. King to Bolling Hall, May, 1824, *Bolling Hall MSS.*

92 *Alabama Republican*, Aug. 10, 1821.

93 *American State Papers, Public Lands*, IV, 805; *Southern Advocate*, Oct. 6, 1826.

94 *Ibid.*

95 *American State Papers, Public Lands*, VI, 188.

son County were invited to attend a meeting for the purpose of regulating the currency of the county. The Merchants' and Planters' Bank of Huntsville accepted Tennessee bank notes only at such a large discount that land sales were being seriously jeopardized. An editorial in the *Alabama Republican* of October 6 clearly set forth the state of affairs:

> We cannot at this time pay for our lands which are the largest portion of our debts with Tennessee paper. The only things we can purchase with it are exchanges and country produce at an advance price, fully equal to the depreciation. . . . The fall of the Fayettville Bank has caused us to distrust the other banks in that state; for who can examine the causes which led to that event and not declare that the Nashville and other subordinate banks will follow its example at some not so distant period. Another reason for rejecting their paper is, the state of Tennessee is upon the eve of flooding us with another emission of $300,000 of that vile trash without one cent to redress it. Shall we quietly wait until this enormous quantity is located firmly upon us and then say that we will regulate it? No sir, the state of Alabama ought to rise as one man and expel this ghost of money from its limits. I affirm that the seven counties of the Big Bend are the supporters of this fiat money and that the state of Tennessee, with our aid, is in a state of bankruptcy.

The facility with which banks granted loans gave an unnatural stimulus to the purchase of public lands and to the extension of the cotton industry. Settlers took up large tracts of land, mortgaging their negroes to the banks for loans with which to carry on the planting industry, and then turned over the cotton as payment on the loan. Trusting in the high price of cotton, the banks frequently advanced far beyond the amount that wisdom dictated and were thus forced into insolvency. In 1825 the inevitable happened in Madison County when the "Huntsville bank departed this life on the 5th instant, covered with dishonor. She died as she had lived, beloved by her friends and detested by all the rest of the world."[96] Just before closing, the bank flooded the vicinity with worthless paper, even though specie from fall cotton sales had been deposited in considerable amounts. However, by appropriating the difference in value between good money and their own paper, the owners were supposed to have realized between $200,000 and $300,000. A great cry arose that "the operators ought to be held up to the indignation of a defrauded public", but it was merely a case of the old, old story.[97]

Sentiment for preëmption rights was strong in Alabama. The Acts of 1830 and 1831 allowed purchasers who had taken further credit, and who were unable to hold their lands, to preëmpt the forfeited tracts at from $1.25 to $3.50 an acre. For those who still occupied their relinquished lands the act granted a similar preëmption right. In such cases the planter got rid of the poorest of his holdings and tried by means of the new credit basis, generally for eight years, to keep the choicest parts of his plantation. These lands had been bought at prices ranging up to thirty dollars and over per acre. In order to meet such unreasonable prices the purchaser could not bring these new lands into profitable production by the time payments were due. Therefore, much land began to revert to the government by 1829,[98] but by the Acts of 1830 and 1831 these lands could be preëmpted at not over $3.50 an acre including former payments, or $1.25 per acre if originally purchased at fourteen dollars or less.[99] In this way a considerable quantity of high-priced lands in Alabama passed into private hands at only nominal figures.

The French Association of Marengo County claimed

---

96 *The Huntsville Democrat,* Feb. 15, 1825.

97 *Ibid.,* Jan. 11, 1825.

98 *Mobile Commercial Register,* Feb. 13, 1829.

99 *Southern Advocate,* April 9, 1830; Treat, *The National Land System,* pp. 158-159.

the right of preëmption under this act. Prior to October, 1831, all persons who had settled and cultivated the land could enter a quarter section for $1.25 per acre provided it did not conflict with the French grant. In July, 1833, A. J. Crawford, the Register at Demopolis, wrote the Commissioner of Public Lands to learn how receipts must be issued in order to safeguard the French claims.[100]

It was not until 1836 that a bill was passed by Congress to reduce the price and prohibit land sales by public auction. When two or more people applied to buy, preference was given to the occupant. No amount of land larger than a section could be bought at one time, and land could be bought only for purposes of cultivation and settlement. By this law tracts were required to be occupied three years after purchase before a patent could be obtained and the price was reduced to a minimum of $1.25 per acre. The main idea of the law was to stop entries of public lands for purposes of speculation.[101]

The final removal of the Indians in 1836 threw open for settlement a large area in Eastern Alabama, and once again immigrants rushed in to take up new lands. The rise in land prices brought about by this movement caused many planters from Georgia and the Carolinas to pass on through Alabama and settle in Mississippi, Louisiana and Texas. A North Alabama newspaper stated, "In truth Alabamians who wish to invest all their capital in negroes and can spare nothing for lands, should go to Mississippi, Louisiana and Texas. To this land of promise, thousands are wending their way. . . ." The movement was especially noticeable in the towns along the Alabama River, where travelers took passage to Mobile on their way westward.[102]

Although even the founding of the State banking system failed to check the evils of unlimited loans secured,

---

100 *Letter Book of the Demopolis Land Office*, 1833-1856.

101 *American State Papers, Public Lands*, VIII, 883.

102 *The Huntsville Democrat*, May 27, 1835.

in many instances, by the proceeds of the next year's crop, the Panic of 1837 brought to an end the mania for buying lands and slaves fostered by the employment of credit in the State banks.[103] Before the Panic, the banks lent money lavishly on land, cotton and slaves at inflated values. As Jackson stated:

> The receipts from public lands were nothing more than credits on the bank. The banks let out their notes to speculators which were paid to the receivers and immediately returned to the banks to be sent out again and again. Thus they were means of transferring to the speculator the most valuable part of the public domain.[104]

After the Panic of 1837, during which a great portion of the financial loss was borne by the commission merchants and business men of Mobile, the State Bank, in order to obtain more specie, determined to enter the cotton market.[105] Agencies were organized, post notes issued, and advances made to planters on crops which were to be shipped abroad in the hope that bills of exchange would bring in specie. Thus it was expected that the planter would have more money at his disposal and would be encouraged to raise cotton. The scheme was plausible enough, and for a time gave fair promise of realizing the cherished object. Newspapers contained flaming advertisements of "Cotton Agents" and "Agents for Advancing on Cotton", post notes were issued, as well as ordinary notes of the banks, and warehouse certificates were prepared by the thousands.[106]

In 1841 a law was passed that advances could not exceed 25% of the anticipated value of the crop and that a bill secured by two endorsers must be given, payable in

---

[103] William O. Scroggs, "Pioneer Banking in Alabama", *Facts and Factors in Economic History*, pp. 416-420.

[104] Lewis C. Gray, *History of Agriculture in the South to 1860*, II, 899; Reginald Charles McGrane, *The Panic of 1837*, p. 44.

[105] William Garrett, *Reminiscences of Public Men*, p. 63.

[106] *Selma Free Press*, Aug. 8, 1839.

nine months at 6% interest. Cotton was received by the State Bank and shipped at the risk of the owner, who might limit the price and date at which the cotton could be sold. However, this power ceased at the expiration of four months after the cotton arrived at a foreign port.[107] Upon its sale, the net proceeds were to be placed to the credit of the bill with interest and the excess, if any, paid by the bank to the shipper, who was also to be entitled to the benefit of the difference in exchange when the cotton was sold in a foreign port. The bank was to be responsible for no losses except those which arose from the misconduct of its agents. In the event the proceeds of the cotton did not take care of the bill, the bank agreed to make good the deficiency.[108]

Under the existing laws it was almost impossible to control speculation, which invariably returned when prices were high. Planters borrowed far beyond their means to pay in order to buy land and slaves, which brought financial chaos to the banks as soon as the market slumped. The State legislature held numerous investigations but apparently was unable to stem the tide of over-speculation.[109] In almost every case the directors of the State Bank or its branches failed to use sound financial judgment in making loans secured by future crops.[110]

Recovery from the Panic of 1837 was comparatively rapid, but a depressed cotton market and low prices in the early forties tended to hold back the price of land and slaves. In 1850, however, lands which formerly sold for $1.25 per acre brought from five to ten dollars. In the same manner the average price of slaves rose from $600 to $1,000, thus tending to increase the cost of pro-

[107] The Bank of the State of Alabama vs. J. M. Bates, *The Plea and the Opinions of the Judges of the Supreme Court of Alabama*, p. 17.

[108] *Ibid.*, p. 18.

[109] *Report of the Commissioners to Examine the Bank at Mobile*, Oct., 1855.

[110] *William P. Gould Papers*, Feb., 1845.

duction. If the planter held his cotton until the manufacturer's stock became low he could usually receive a better price, but a more accurate knowledge of market conditions enabled the manufacturer to bargain with the planter if he was forced to do so.[111] Until 1860 the price of land and slaves depended largely on the price of cotton, but at this point the cost of negroes forged far ahead. Black Belt lands sold at from fifty to eighty dollars an acre, while the price of prime field hands increased to $1,200 or $1,600, yet cotton fluctuated at between ten and eleven cents.[112]

In 1820 slaveholding was restricted to a limited class, for a large percentage of the immigrants who had come in from the older states owned no slaves. In Limestone County, Thomas Bibb, Joseph Peoples, Nicholas Davis, Benjamin Fox and James Jackson were the only planters who owned more than twenty,[113] and their yields of cotton during that year averaged about forty bales.[114] During the next decade, however, the high price of cotton and frugal management enabled many who had formerly owned none to purchase slaves, while at the same time planters who were already slave owners busied themselves to increase their own holdings. This perhaps may account for the 130% increase in slaves in Limestone County by 1830.[115]

During the same period the Black Belt counties, though showing an increase in slave population,[116] were not considered well adapted for cotton culture. However, this idea was changed in the early thirties when an upland variety was introduced which was known to be blight proof when grown in limestone soil. Before 1830 the

---

111 *DeBow's Review*, Jan., 1852, p. 80.

112 *DuBose Canebrake MSS.; Alabama Beacon*, June 29, 1860.

113 Bibb, with 110 negroes, was probably the largest slave owner in the State, *Manuscript Census Returns*, Limestone County, 1820.

114 *Ibid.*

115 *Ibid.*

116 *Fifth Census of the United States*, p. 2.

average size holding in Dallas and Wilcox Counties probably amounted to no more than six or eight slaves,[117] though James and Charles Tait, George Williamson and John Eads had approximately fifty each.[118]

In Franklin County, which had 1,676 slaves as compared with 3,312 free whites, only seven people owned more than seventeen slaves, while the average size holding was one or two.[119] In Shelby and St. Clair Counties there were few holdings as large as ten, the greatest majority owning no slaves at all.[120] The total increase in the entire State between 1820 and 1830 was 74.5%.[121]

The increase of slaves in Alabama was followed closely by an increase in the production and export of cotton. Crop estimates in 1820 were given at 25,390 bales, while six years later they had increased to 120,000.[122] In the same way the value of exports through Mobile had increased from $96,000 in 1820 to $2,294,594 in 1830. This, of course, did not include North Alabama cotton, which was shipped to foreign markets through the port of New Orleans and amounted to almost two million dollars.[123]

High prices and cheap lands caused Alabamians to be completely absorbed in the production of cotton.[124] In 1828 a traveler recorded that on his way from Montgomery to Mobile the boat stopped at about twenty places to take on bales of cotton, and at each landing place people came on board to discuss the prospects of the market.

---

117 *Manuscript Census Returns*, Wilcox County, 1820.

118 *James A. Tait Plantation Memorandum Book*; *Manuscript Census Returns*, Wilcox County, 1820.

119 *Manuscript Census Returns*, Franklin County, 1820.

120 *Manuscript Census Returns*, Shelby County, St. Clair County, 1820.

121 *Fifth Census of the United States*, p. 2.

122 *Niles Register*, XXIV, 21; XXX, 156.

123 *The Huntsville Democrat*, July 4, 1833.

124 Part of the French Grant in Marengo County was opened for sale in 1830, *Greene County Gazette*, Sept. 6, 1830.

> Every flaw of wind from the shore wafted off the smell of that useful plant; at every dock or wharf we encountered it in huge piles or pyramids of bales and our decks were soon choked up with it. All day, and almost all night long, the captain, pilot, crew and passengers were talking of nothing else; and sometimes our ears were so wearied with the sound of cotton! cotton! cotton! that we gladly hailed a fresh inundation of company in the hopes of some change — but alas! Wiggin's Landing, or Choctaw Creek, or the towns of Gains or Cahawba or Canton, produced us nothing but fresh importations of the raw material. 'What's cotton at?' was the first eager inquiry. 'Ten cents! Oh that will never do!' From the cotton in the market they went to crops in the fields — the frost which had nipped their shoots — the bad times — the overtrading — and so round to prices and prospects again and again, till I wished all the cotton in the country was at the bottom of the Alabama![125]

The forties witnessed the expansion of the cotton kingdom into the Black Belt of Alabama in much greater proportion than its subsequent increase in the northern part of the State. The counties of Dallas, Marengo, Greene, Montgomery, Perry, Lowndes and Sumter showed gains in slave population far in excess of other parts of the State. The largest single increase was in Marengo County, which had 2,987 slaves in 1830 and 11,898 in 1840. Dallas, however, had amassed the greatest slave population, with 17,207 slaves within its borders. Roughly speaking, the total gain for the above-mentioned counties had been greater than 200%.[126] It is worthy of note that the census of 1830 shows that the proportion of blacks and whites was about equal, whereas by 1840 the increase in slaves had greatly outstripped that of the free whites.[127] In the typically nonslaveholding counties of North Alabama the reverse was the case, for blacks tended to increase very little, while the whites made sub-

---

125 Basil Hall, *Travels in North America*, III, 310-311.

126 *Sixth Census of the United States*, p. 1.

127 *Ibid.*

stantial gains. It appears that in the Black Belt the great increase in negro population came not so much from additions to the slave holding class as from expansion of individual holdings by purchase and natural increase.[128]

The cotton crop was checked rather sharply in 1838 and 1845, due in the first case to an early frost which came on October 7, and in the second to ravages of the caterpillar.[129] In August of 1844 the caterpillar first appeared on the river plantations below Claiborne, but its full effects did not reach Central Alabama until 1847, when the cotton crop of William P. Gould fell off from an average of 150 bales a year to seventy-seven.[130] Four years before, the cotton bollworm[131] had ravaged the Tennessee Valley, and so great was the distress among the planters that William H. Gee of Madison County wrote his son, James, who was then a student at the University of Virginia, that "our cotton crops will be very short in consequence of the worm. You must be very prudent and not spend a dollar more than you are compelled as I shall have a use for every dollar I can raise and I am afraid more than I shall get for my crops."[132]

In 1850 began the era of the general rounding out of the cotton kingdom in Alabama. The upper Coosa Valley, heretofore uninvaded by the plantation system, came into prominence as a cotton-producing area. From this region cotton was shipped by river to Rome and thence by way of the Georgia railroads to Charleston and Savannah. A typical plantation in this part of the State was that of Benjamin C. Yancey, who in 1851 acquired 784 acres in Cherokee County at a total cost of $10,357, and

---

128 *Ibid.*

129 *DeBow's Review*, XVII, 429.

130 *The Southern Cultivator*, II, 197; *William P. Gould Papers.*

131 Not to be confused with the modern boll weevil.

132 *Gee MSS.*, Nov. 18, 1840.

immediately began the culture of cotton with sixty-five slaves.[133]

In general, the decade was a period in which planters bought up the holdings of their poorer neighbors, as may be shown by the loss of white population in both Clarke and Madison Counties.[134] However, this tendency to consolidate has been greatly exaggerated, since the planter and nonslaveholder were found living side by side in every county of the State. For instance, in Dallas County, which is typical of the Black Belt and which in 1850 had a population of 7,014 whites and 20,925 negroes, the manuscript census returns indicate that only 510 heads of families owned as many as ten slaves. Of this number 158 owned thirty or more, 100 heads of families owned at least fifty slaves, and thirty-five owned 100 or more. William P. Molett was the largest slave owner in the county with 307.[135] In Perry County, of 9,425 whites, only 1,081 heads of families owned any slaves and of this number 419 owned more than ten.[136]. In Sumter, a rich planting county in the extreme western part of the State, the same condition seems to have existed.[137]

Although acreages varied in size from fifteen to 9,000,[138] about half of which was usually unimproved lands, the census lists many small plots of from forty to eighty acres. The agricultural schedule in Lowndes County indicates that from one to ten bales of cotton were more common than larger yields. In 1850, only

---

133 *B. C. Yancey Coosa River Plantation Account Book.*

134 *Seventh Census of the United States* (Population); *Clarke County Democrat*, Jan. 31, 1856.

135 *Manuscript Census Returns*, Dallas County, 1850.

136 *Manuscript Census Returns*, Perry County, 1850; also see returns for Greene, Autauga, Macon, Tuscaloosa, Wilcox, Montgomery and Marengo Counties.

137 *Manuscript Census Returns*, Sumter County, 1850.

138 Six plantations of James A. Tait in Wilcox County. (He also owned a 600-acre plantation in the Mississippi Delta.) *James A. Tait Plantation Memorandum Book.*

thirty-one planters seem to have produced 100 or more bales, while the largest single production was 344.[139] In the same year 1,429 people listed their occupation as that of planter and 106 were listed as overseers. This might tend to show that of the number who classed themselves as planters, only a fourteenth had holdings large enough to employ an overseer.[140] The William P. Gould plantation in Greene County was typical of the medium-sized holding. Gould owned fifty slaves and over a period of years produced from 120 to 175 bales of cotton annually on 1,900 acres of land, the greater part of which was unimproved.[141]

In the hill counties of Central and North Alabama holdings were much smaller than in the Black Belt. In Blount County there were ninety-nine slave owners, but the greatest number of slaves belonging to any one person was twenty.[142] In Cherokee County there were 345 owners but holdings averaged only three or four slaves. The same was also true for Coosa, Marshall, Randolph and other counties of this region.[143]

From the tax list of Lawrence County in 1860 one is able to ascertain that there were about 930 persons who paid taxes. Of these, 407 owned slaves and 523 were nonslaveholders. Slave owners who apparently owned no land amounted to eighty, and one of this number owned eighty-five slaves.[144] Further observation indicates that there were seventy-eight non-slave owners who owned from 100 to 200 acres of land, twenty-one who owned from 200 to 300, eleven 300 to 400, and eight over 600 acres. From this one might infer that certain slave-

139 *Manuscript Census Returns*, Lowndes County, 1850.

140 *Ibid.*

141 *William P. Gould Papers.* (See Plantation Diaries, 1840-1860.)

142 *Manuscript Census Returns*, Blount County, 1850.

143 See *Manuscript Census Returns* for Cherokee, Coosa, Marshall and Randolph Counties, 1850.

144 *Tax List*, Lawrence County, 1860.

holders rented from their nonslaveholding neighbors, who found it profitable to lease their surplus lands.[145]

Other things indicate that the nonslaveholder was in many cases a rather hard-working, well-respected individual. County records show that many times he left considerable property in the form of lands, stock, agricultural produce and household furnishings. More interesting still is the fact that promissory notes, usually in small denominations but sometimes totaling several hundred dollars, are found probated as assets. Possibly these represent loans to planters, who in times of adversity called on their supposedly less prosperous neighbors for supplies of corn and pork.[146]

In general, planters of the Black Belt preferred to reside in small communities rather than on their plantations, which were frequently in the prairies or river bottoms and extremely unhealthful in summer. Such communities as Dayton, Tilden, Greensboro and Mount Meigs were typical settlements of the planting aristocracy.[147] Nearly every plantation on the river had its own landing, which, until the coming of the railroads in the fifties, served as a depot from which cotton was shipped to Mobile and plantation supplies were received from merchants in the city.

The period from 1850 to 1860 was the decade of greatest cotton production in the ante-bellum South. At the beginning of that ten-year era Alabama had risen to first place among the cotton-producing states of the country with a peak production of 564,429 bales.[148] The process

---

[145] *Ibid.*; A system of tenant farming evidently existed in Monroe County, for in 1850 the census taker listed 150 tenants, *Manuscript Census Returns*, Monroe County, 1850.

[146] *Probate Record*, Greene County, Vol. M; Dallas County, Vol. M; Coosa County, *Will Book* A.

[147] *DuBose Canebrake Manuscripts*; Frederick L. Olmsted, *A Journey in the Seaboard Slave States*, p. 575.

[148] *Seventh Census of the United States* (Agriculture), pp. 429-431.

of consolidation continued uninterrupted, as is indicated by the increased average number of acres in each farm, which in 1850 was 289 and in 1860 was 346.[149] In cash value, the Black Belt plantations were estimated far above those in the rest of the State. The counties of Dallas, Greene and Marengo, alone, contained agricultural properties estimated at nearly twelve millions, while in the Tennessee Valley Madison County contained over three millions.[150] Dallas County, in the center of the Black Belt, had the largest slave population in 1860 with 25,760, whereas Chambers County in the Piedmont with less than half as many slaves had more slave owners than any county in the State. The census of 1860 shows that there were 33,730 slave owners in the State who owned 435,080 slaves, or an average of 12.8 slaves per slave owner. Likewise, in comparison with the total white population of the State, which was 526,431, slave owners constituted only 6.4%.[151] Considering five as the average number of persons per family, these figures indicate that slightly less than one-third of the white population in the State belonged to the slaveholding class.

As early as 1846, Sir Charles Lyell noted that Alabamians were leaving the State to take up lands in Texas. At Claiborne he encountered a group from the Tennessee Valley with forty slaves loaded in heavy wagons about to launch forth on an expedition to Texas and Arkansas.[152] A few days later, on a boat between Mobile and New Orleans he made another observation concerning this emigration movement westward:

> On board were many 'movers', going to Texas with their slaves. One of them confessed to me, that he had been eaten out of Alabama by his negroes. He had no idea where he was going but after settling his family at Hous-

---

149 *Eighth Census of the United States* (Agriculture), p. 222.

150 *Ibid.*, pp. 429-431.

151 *Ibid.*, p. 223.

152 Charles Lyell, *A Second Visit to the United States*, II, 56.

> ton, he said he should look out for a square league of good land to be had cheap. Another passenger had, a few weeks before, returned from Texas, much disappointed, and was holding forth in disparagement of the country for its want of wood and water, declaring that none could thrive there, unless they came from the prairies of Illinois, and were inured to such privations.[153]

During the fifties this movement out of Alabama was even more accentuated, although the population of the State continued to grow.[154] In 1857 the *Jacksonville Republican*, quoting a Memphis newspaper, stated:

> We have never before observed so large a number of Emigrants going westward as are crossing the river at this point daily. The two ferry boats (sometimes three) are crowded from early morn until the boats cease making their trips at night. It is no uncommon sight to see from twenty to forty wagons encamped on the bluff for the night, not with standing there has been a steady stream going across the river all day. . . . The emigration is from the older states, mostly Virginia, North Carolina, South Carolina, Georgia and Alabama.[155]

In addition to the exodus of planters to Texas and Arkansas, the Kansas fever seized many Alabamians. Emigrant companies were organized throughout many parts of the State, following the lead of the Buford expedition which, with 200 persons, left Montgomery in 1856.[156] On the eve of secession a group of Alabama emigrants left the State to settle in the Gila valley in Arizona. They expected to follow the route of the overland mail to El Paso, where the party was to be joined by others from the North, and then proceed westward together.[157]

The expansion and growth of the cotton kingdom in Alabama followed somewhat the trend of the other states

---

153 *Ibid.*, p. 88.

154 See *Eighth Census of the United States* (Population), p. 8.

155 *Jacksonville Republican*, March 11, 1857.

156 *Montgomery Advertiser*, April 9, 1856.

157 *Clarke County Democrat*, April 14, 1859.

in the Southwest. The years between 1820 and 1860 witnessed the spread of the plantation system from the river valleys into the Black Belt and hill country, and the consequent shifting of population from one region to another. By 1860 the slave population of Dallas County had increased over 900%, while in Blount, a typical small slaveholding county, the gain was slightly over 300%. In the same two counties the change in white population during the same period showed an increase of from 3,000 to 7,000 in Dallas and from 2,000 to 10,000 in Blount. By 1860 the average size farm in the Black Belt county was 300 acres and in the hill county thirty-five, which is an indication that the consolidation of holdings was much more accentuated in the central part of the State than in the hilly regions of the north. The geographical limits of slavery had become rather definitely fixed by 1850, and Alabamians were turning to Texas and Arkansas in search of new lands.

CHAPTER III

PLANTATION MANAGEMENT

ONE of the important personages in the operation of the plantation was the overseer. "Unnoticed in society, with no friends to record his services, he lived and disappeared without leaving a record of his existence."[1]

The agreement between the overseer and his employer was usually in the form of a contract, sometimes verbal but much more often written. Typical is the contract made in 1860 between W. P. Bocock of Marengo County and his overseer, Lewis O. Collins. The agreement was to run for one year, and Collins was to give his whole time, attention and skill to carrying on the business of the plantation. Above all, he had to observe the wishes and directions of his employer when they were made known, being duly careful of the good conduct, health and cleanliness of the negroes, to care for the sick, and to see that the slaves complied in a reasonable manner with the rules laid down for them. Also, he was to conduct himself with prudence, sobriety and a faithful regard for his employer's interests. Bocock agreed to furnish him with a horse to ride about the plantation and a negro woman to cook and wash for him. Collins' wages were to be $500 for twelve months, but Bocock reserved the right to end the agreement at any time he was not satisfied, provided he paid him his wages up to date.[2]

---

[1] John Spenser Bassett, *The Southern Plantation Overseer as Revealed in his Letters*, p. 1. For other good general descriptions see Ulrich B. Phillips, *Life and Labor in the Old South*, Ch. XV; R. B. Flanders, *Plantation Slavery in Georgia*, Ch. VI.

[2] *Contract between W. P. Bocock and Lewis O. Collins*, Marengo County, 1860.

In many instances a list of regulations such as the following accompanied the contract:

1. Humanity, as well as policy would dictate that my negroes will be well fed and clothed and comfortably lodged. To the negroes of this plantation rations of meat are issued monthly and on a stated day of each week, one peck of meal per each grown hand. Clothing and building repairing will be attended to as occasion may require.

2. It will be required of the overseer at all times to give his individual attention to the business of the plantation and as far as practicable to be with the hands in the fields. If at any time it should be necessary for him to leave the plantation, it is expected that he will notify the proprietor of the same, that he may take charge of the hands during his absence. If particular directions are given or suggestions made by the proprietor to the overseer in relation to the affairs of the plantation it is expected that they will immediately be attended to; but in order that nothing may be neglected, he will in all other cases use his own judgment and discretion. In order, to as much as possible, economize time, and that the attention of the overseer may not be taken from the hands while in the field, it is desirable that a part of such jobs as will admit of delay, may be reserved for wet weather; and when the old women cannot be advantageously employed out of doors, they will be required to perform the stated task of spinning.

3. The mode of feeding the work horses will be directed as the overseer may think best, but it is expected that he will assure himself that it is regularly and carefully attended to, and that they have salt at least once a week, particularly during the plowing season. It is expected that he will from time to time count the stock of sheep and hogs and that he will make some one accountable for their general superintendence.

4. No negroes belonging to the plantation will be allowed to leave it without a pass from the proprietor, and no one from the neighborhood or elsewhere will be permitted to frequent it without written permission to do so. In order to ascertain if this rule is observed, it will be required of the overseer to inspect the cabins at different hours of the night as often as once in every week.

5. No negro will be suffered under any pretense whatever to strike another — excepting always such moderate

correction as it may be necessary for parents to use with their children. They will not be permitted to have whiskey or other intoxicating spirits in their cabins — the drams that it may occasionally be thought proper to give them to be drunk immediately and not suffered to accumulate.

6. If a negro reports sick he or she will be sent immediately to the dwelling house for examination and when considered by the proprietor to be again fit for duty it will be so reported to the overseer.

7. When it becomes necessary to punish a negro for neglect of duty, disobedience of orders or other offense, the punishment is inflicted cooly [sic] and dispassionately, with a whip to be kept for that purpose — Beating with a stick, the fist, or kicking are positively forbidden. Care is to be taken to proportion as far as may be, the correction to the offense, but in no case to desist until submission and penitence are manifested.

8. The overseer will eat at the dwelling house, or his meals will be furnished at his own room, as may from time to time be most convenient to the family. The engagement between the proprietor and the overseer may be dissolved, whenever either party may desire it, and payment will be made for the time that service has been rendered.

December 23, 1837.

Ludwig Henderson has this day engaged with me after having read and agreed to the foregoing.

William P. Gould and L. Henderson

(Pay $500 per annum)[3]

Lists of instructions to overseers were many times more detailed than the above. Charles Crommelin set down twenty-eight separate items, among which were:

He [the overseer] must himself keep the keys of the cribs, smoke house and all other buildings in which any property belonging to me is stored, he must have poultry raised and well taken care of that I may be furnished with it as required; he must have special care and attention given to the dairy, seeing that the milch cows are regularly fed, salted, and milked and that butter is properly and neatly prepared and sent to me regularly, and buttermilk when desired.[4]

---

[3] *William P. Gould Papers.*

[4] *Crommelin Papers,* 1845.

Both Crommelin and Colonel Thornton, owner of the "Thorn Hill" plantation, required that their overseers go about their plantations on foot, believing that negroes could not be properly supervised from the saddle.[5]

In was generally felt that every planter should set down in writing distinctly and emphatically the duties he wished his overseer to perform. However, it is very probable that the overseer did not follow the terms of this instrument exactly, nor did all employers attempt to have so complete a body of rules put into operation. The contract is interesting chiefly because it indicates the high standard for the conduct of a good overseer.[6]

Olmsted found that the salaries of Alabama overseers averaged from $200 to $600 per year. An exceptional man perhaps received as much as $1000 and on rare occasions from $1,200 to $1,500.[7] Usually the high salaries went to men in charge of plantations of absentee owners, where superior ability was a primary requirement. In 1841 C. C. Clay, Jr., of Madison County wrote his father saying, "I have employed a man by the name of John Tanner to attend to the Oakley place and have agreed to pay him $185. He comes to me very highly recommended for industry, sobriety and honesty — which last quality I shall only credit when it is proven by trial."[8] For his numerous holdings and large number of slaves, James Tait in 1851 employed three overseers. His plantation memorandum book records the following information: "Engaged Brown for the Home Place at $375, Newton for the Old Place at $250, and Lewis for the Lower Place at $275."

---

[5] *Ibid.*; *Cotton Book of the Thorn Hill Plantation*, 1856. For further rules for overseers see *B. C. Yancey Coosa River Plantation Account Book*, 1858; *The Southern Cultivator*, V, 62.

[6] *American Cotton Planter*, II, 372; Bassett, *The Southern Plantation Overseer*, p. 24.

[7] Frederick L. Olmsted, *A Journey in the Back Country*, p. 57.

[8] C. C. Clay, Jr., to C. C. Clay, Sr., Feb. 6, 1841, *Clay MSS.*

The tenure of an overseer was usually short-lived. James Tait changed overseers nearly every year, and his agent in 1837 was particularly annoying. Among his offenses were failure to tan leather, neglecting a fire in the field and permitting a fence to burn, neglecting livestock and slaves, which resulted in the death of two horses, a mule, a steer and two slaves; allowing the negroes to cut fence rails on a neighbor's land; causing the negro men to run away by interfering with their women, and keeping the neighbors at odds with Tait by harmful gossip.[9] In 1835 Charles Tait recorded, "A legacy to my children. Never employ an overseer who will equalize himself with negro women. Besides the immorality of it there are evils too numerous to be mentioned. My maxim is, never to have any trading or dealing with an overseer."[10] Two years later the following was recorded: "Never, never talk to overseers about your neighbors; they are mischief makers."[11]

Other Alabama planters also had overseer troubles. William P. Gould discharged Naugher for striking a negro woman with an ear of corn.[12] His neighbor, Colonel Thornton, got rid of two overseers in three years for neglect of duty.[13] One of the most flagrant instances of neglect of duty was that of the overseer of Charles Mathews on his plantation in Wilcox County. On returning from a six weeks' business trip, Mathews found that the overseer had not been to the field during his absence. The entire cotton crop was ruined and the negroes had all become sick from neglect.[14] Frequently the rapid turnover resulted in overseer shortages. News-

---

9 *Charles Tait Journal.*

10 *Ibid.*

11 *Ibid.*

12 *William P. Gould Papers*, Diary, Jan., 1855.

13 *Cotton Book of the Thorn Hill Plantation*, 1855-1857.

14 Charles Mathews to George Mathews, April 13, 1834, *Frank Mathews Papers.*

paper advertisements such as the following were quite common:

> Wanted a young man of good character, of industrious habits and if not personally known to me, with certificates to show that he is moral, industrious, and tolerably intelligent, to act as foreman on my farm — that is to act entirely under my direction. I would have advertised for an overseer, but I have never been able to employ one that I was pleased with, when he has had previous experience.[15]

In the early days of the plantation era in Alabama, the general practice was to select an overseer merely on the basis of experience and sobriety and look to the number of bales of cotton he produced as a test of good management.[16] However, with a depressed cotton market, worn-out lands, rising price of slaves, and keen competition in general, other qualifications were necessary. To the average overseer it was of no consequence if old hands were worked down, young ones overstrained, if breeding women miscarried, mules were broken down, stock neglected, plantation tools destroyed and lands ruined, so long as the requisite number of bales was made. Frequently the overseer was allowed to purchase supplies on his own order without accurate bookkeeping. The planter put up with such inefficiencies merely because a good crop was made, but once the overseer was allowed to take such liberties the planter was doomed to failure.[17]

In general, overseers were violently opposed to anything new in agriculture. Ignorance, for the most part, prevented them from learning the principles of scientific farming.[18] The following letter from the overseer of Samuel Pickens is a typical example of ignorance:

---

[15] *Alabama Beacon,* Dec. 1, 1849.

[16] *American Cotton Planter,* II, 356.

[17] *The Southern Cultivator,* II, 107.

[18] *Ibid.,* II, 108.

gooden Place

Dear frind with Pleasure I seat mi self to drop you afew lines to let you now how we argitenalong her we argitenalong very well now with the cotton. . . . I sent 20 bushel of meal to Greensburer pay day I went up mi self . . . but hit waren sellen so I put hit down to 75 cence I wish you to send me some pills and Some Caster oil that Rubarb hant got no strength in hit. their's Some of the niggers Sick thom [Tom] was sick when you was down here has picked over his 100 pounds ever day you had better come down as soon as you can I think you had better send day [Day] to work on the Ware house for I have got the timber reddy for him what I lack I can git befor he will nead hit give mi best complemence to Miss Pickens and pompy [Pompey]

yourse respectpus

J. Wooten[19]

By 1850 the inefficiencies of overseers were beginning to gain a prominent place in the Southern agricultural magazines. The editor of *The Southern Cultivator* stated in 1848 that "one of the biggest objections to employing negro slavery in this country is involved in the necessity, real or apparent, of employing overseers".[20] He seemed to be of the opinion that no one should own slaves who was not his own overseer or who did not give steady and systematic attention to the man he employed. "The destruction and waste of property committed to his charge by the carelessness of an overseer, the positive mischief of the indulgence of his evil passions, and the failure of crops occasioned by his ignorance and remissness are some of the principal sources of loss which absent proprietors are often doomed to encounter."[21]

In answer to these charges of inefficiency, an overseer from Burnsville, Alabama, replied in the *American Cotton Planter* of 1854. He prefaced his remarks by stating

---

[19] *Samuel Pickens Papers.*

[20] *The Southern Cultivator*, VI, 134; *Selma Free Press*, Nov. 21, 1839.

[21] *The Southern Cultivator*, VI, 134-135.

that overseers were never given a fair trial, that crops went bad and agents were discharged. By the very nature of their work overseers had to have authority, and no man who had to be ordered what to do was fit for the job. If this were the case negro drivers would be more desirable, for, generally speaking, planters had more confidence in their negroes than in their overseers. "If they wish to know how their business progresses they seldom, if ever inquire of the white man, but call upon some negro and ask him questions that ought to make any gentleman blush to think of asking a negro about a white man." The author further stated that with so little confidence manifested in a man it was no wonder that he felt little interest in raising negro children, enriching lands and improving the stock. The fact that every overseer was on trial during the first year was also against him. Few men could prove their talents as managers when everything was new, before they had an opportunity to learn the dispositions of their employers, abilities of the negroes, and the quality of the land. If the planters had changed their policies in this respect, their overseers would probably have taken more interest. In some instances, planters themselves had very poor judgment about running plantations and instead of listening to the advice of their overseers, they were prone to adhere to the suggestions of their neighbors. However, at the same time they placed the blame on the men they held responsible.[22]

In most cases, overseers in Alabama were sons of the piney woods folk who sought to escape the struggle for existence on poor soil. Here and there an exceptional man, by hard work and frugal living, would rise above his fellows. Such a man was John Catlin, who had been an overseer for Wade Hampton on one of his Mississippi plantations. Catlin saved his money and bought land in Marengo County, and by starting on a small scale with only a few

[22] *American Cotton Planter*, II, 150.

slaves, gradually increased his holdings until he became a prosperous planter.[23]

On entering the employment of a planter an overseer's first duty was to inform himself of the condition of the plantation, stock, negroes and implements. His next responsibility was to learn the views of his employer as to the general course of management, and if any differences of opinion occurred, a wise overseer soon reached a definite understanding with his employer. His first consideration was the care of the slaves, stock and plantation equipment. Next he was to see that enough food was produced for use on the place. By food was meant corn, pork, potatoes and vegetables for the slaves and corn, oats, hay and fodder for the stock. These two duties done, he was to raise as much cotton as possible without overworking the negroes.[24]

It was all too commonly the case for masters to look only to the products of their plantations, and measure the worth of their overseers by these standards, without troubling themselves to inquire into the manner in which things were managed. In many instances, the planter lost more in the diminished value of his slaves by overwork than he gained by large crops of cotton. Overseers had no property in slaves and, of course, lacked the check of self-interest. Their only aim was mere crop results, and in this respect masters were often greatly to blame by encouraging such an attitude.[25]

The day's routine on the plantation began at dawn and lasted until sunset. The overseer usually blew a horn to summon the negroes from their cabins. In order to prevent any loitering about, James Tait stated, "To make them get off quickly after the horn is blown, I always

---

[23] *DuBose Canebrake Manuscripts.*

[24] *American Cotton Planter*, II, 353-354; *DeBow's Review*, March, 1855, pp. 339-345; Bassett, *A Southern Plantation Overseer*, pp. 11-12.

[25] *DeBow's Review*, June, 1855, p. 714.

whip the last one out."[26] Tait preferred that his negroes have breakfast before starting for the field, though some planters found it more convenient to grant thirty minutes for breakfast later in the morning. While the negroes were eating it was the duty of the overseer to see that the stock was properly fed and also to visit and prescribe for the sick. The sick were to be visited not only in the early morning, but at such times of the day and night as individual cases might require. These duties having been attended to, the overseer usually repaired to the field, where he remained with the negroes. At twelve o'clock dinner was served. If the hands were within reasonable walking distance they usually returned to the quarters, where after the noon meal they were allowed to rest from one to two hours, depending upon the weather and condition of the crops. Those who were too far away were served in the field, but the same period of rest was allowed.[27] Work was then resumed until sunset, when the hands returned to the quarters. After feeding the stock and putting the equipment in order for the next day's work, a few hours leisure was allowed the negroes, usually until nine o'clock, when the horn was blown and the overseer was to see that every negro was in his cabin.[28]

The plantation diary of William P. Gould for 1852-1853 gives a rather vivid description of daily routine on the "Hill of Howth" plantation in Greene County:[29]

> Jan. 2—Marked and weighed 32 bales averaging 511 lbs.
>
> 4—All hands picking cotton in New Ground except Phil who is hauling fifteen bales of cotton to the river. Hope to make 170 bales this season, but am doubtful.
>
> 6—Killed 22 hogs making 3348 lbs. of meat.

---

26 *James A. Tait Plantation Memorandum Book*, 1852.

27 *Ibid.*, 1841; *DeBow's Review*, Feb., 1853, pp. 177-178.

28 *Ibid.*, Feb., 1853, pp. 177-178.

29 *William P. Gould Papers.*

23—Women and children picking cotton in the field across the branch. David and his gang ginning. Others clearing old field to plant cotton. It laid idle for several years.

30—Bought 20 sheep.

Feb. 19—Phil and Edward had a fight with a knife. Phil is in a serious condition. Edward ran off. Returned in three days.

25—Gave Mr. Gowdey an order on Rives Battle and Co. (Mobile) for $547 for last year's amt. at Gowdy and Elliots store.

27—Paid David Watson $115.71 for pork and beef bought in 1852 and all other claims up to this date.

March 18—Went to Mobile on business stopped at the Battle House, returned on 22nd to Millers Landing.

April 18—Frost killed all cotton above the ground. Corn also injured — both had to be planted over.

August 14—Began to pick cotton in the Gate cut today.

25—Killed a beef for the market.

September 6—Drought worst in years

'Drought, Drought, Drought
East, West, North and South
From all quarters comes the wailing
Of the golden harvest failing'

14—Packed [sic] 17 bales.

November 10—Finished gathering corn.

16—Went to Eutaw and bought shoes for the negroes — Gave them out in the afternoon.

December 23—Gave each hand 7 lbs of fresh beef and 2 lbs of pork. Intended to buy the negroes some whiskey — None in Eutaw — What little available sold for $2.50 per gal.

James A. Tait apparently found it convenient to plan his plantation work well in advance. His plantation memorandum book records the following:

Plantation Business for the Month of January

Finish picking and ginning your cotton, if you are not done by Christmas. Then get your coal wood. Haul out your cotton seed. — Make your plow lines, women to spin thread of nights for ditto. Plow stocks, gear, and tools to be got ready — fences to be repaired — Clearing ground if any to be cleared — logs to be cut up, rolled and burned

> through out the plantation — negro houses to be repaired and *built if wanted.* Always turn over a new leaf and look bright into the business of the coming year. Industry, economy and care, let them be your polar star, your chart and compass. Feed and clothe your negroes well and make them work. Eschew banks and credits. Thank God, I never borrowed money of any bank but once, and that was a small sum. Stand security for no man.[30]

During the forties and fifties great emphasis seems to have been laid by both the agricultural magazines and the newspapers on the division of plantation work by months. *The American Cotton Planter, The Southern Cultivator, Soil of the South,* and other periodicals frequently ran a column outlining work for the current month.[31] It is difficult to determine just how closely these directions were followed in Alabama, as each planter thought his own particular method superior to that of his neighbor.[32]

In distributing work to the negroes, Alabama planters used both the gang system and the task system and frequently a combination of both. James Tait preferred the gang system for plowing and the task system for hoeing. He allotted 300 acres of cotton to a gang of thirty-three hands, which included ten plows, and he found that this force was sufficient to work the whole area once every three weeks.[33] Each member of the hoe gang received a daily assignment of fifty rows of cotton to chop, though when a negro reached fifty years of age his task was reduced to forty rows. Pregnant women and nursing mothers had their tasks reduced to thirty or forty rows and negroes under twenty were given less

---

30 *James A. Tait Plantation Memorandum Book,* 1841.

31 See *The Southern Cultivator,* 1845-1850; *American Cotton Planter,* 1850-1860; *Clarke County Democrat; Alabama Beacon.*

32 See *William P. Gould Papers,* Plantation Diaries, 1835-1860; *Burford Diary*; *James A. Tait Plantation Memorandum Book*; *B. C. Yancey Coosa River Plantation Account Book,* 1850.

33 *James A. Tait Plantation Memorandum Book,* 1850.

than a full task. In the case of negroes in their early teens two or three were usually assigned to one task.[34]

Tait sent negroes to the field at the age of ten, but during the first year they worked with their mothers, together doing one task.[35] One Marengo County planter put his little negroes to work at the age of five, believing that the influence of their mothers was better than being cuffed by other negroes in the capacity of nurses, "which is calculated to make them untruthful and lazy". His policy was to give them jobs which would not overtax their strength and to punish them lightly for insubordination. Reared in this manner they became good workers at the age of twelve.[36]

From June 15 to July 1, when cotton was about knee high, tasks frequently had to be lengthened in order to keep the grass from getting too much of a start. Additional work was always necessary during the picking season, especially if the weather threatened.[37] There were two leisure seasons on the plantation, one a few weeks after the crop was planted and the other a few weeks after it was made. Because of the excessive heat, James Tait found it a good plan to give his negroes a half day on Saturday during June, July and August if they worked reasonably well. After the crop was planted in the spring he granted the plowers half of every other Saturday and at Christmas all hands had three holidays.[38]

The task system had many obvious disadvantages. In the first place, it put a premium on rapid work, because the hand finishing first had an opportunity to choose his next task. This placed a great injustice on the slower worker who did his task more thoroughly, but who was

---

[34] *Charles Tait Journal,* 1840.

[35] *Ibid.*

[36] *American Cotton Planter,* II, 280.

[37] *Charles Tait Journal,* 1840.

[38] *James A. Tait Plantation Memorandum* Book, 1850.

forced to take what was left of the unassigned tasks because of his careful work on the previous ones.[39] In most cases, assignments were made by a negro driver who worked directly under the overseer. The more the driver kept himself away from the field hands the better, for once they felt that he was their equal, the factor of discipline was lost.[40] J. C. Carnothan, overseer for William P. Gould, wrote that "the negroes work verry [sic] well when I am in the field, but when I am gone Stewart [Gould's driver] will not make them work. I have tried talking to him but it does no good."[41]

On the larger plantations frail negroes were often assigned to spinning, weaving, sewing or household work, while slaves of more than average abilities were employed altogether as carpenters, blacksmiths, cobblers or on other specialized forms of work. In such cases the task system may or may not have been used. On the Tait plantation women in the sewing room were assigned two pairs of pants or two shirts to be made each day.[42] There was always a supply of domestic servants in each well-to-do planter's household, and the occasion frequently arose for detaching part of the field force for other tasks about the plantation. On most estates, laborers were divided into two groups for the better adjustment of the slaves to the requirements of their tasks. It was often customary to set apart the strongest laborers as plow hands while the rest used the hoe, each group working under its own supervisor. On the Burford place practically all ditching, clearing of new ground, chopping of cotton stalks and picking was done by women, while men were employed in plowing, hauling, repairing fences and splitting rails.[43]

---

39 *The Southern Cultivator*, IV, 44.

40 *The Farmers' Register*, IV, 115.

41 Carnothan to Gould, May 26, 1847, *William P. Gould Papers.*

42 *James A. Tait Plantation Memorandum Book*, 1850.

43 *Burford Diary*, 1850.

In managing negroes most Alabama planters were of the opinion that firmness of manner and promptness to enforce obedience saved trouble and prevented excessive whipping. If the negro felt that the master was his lawgiver and judge, but at the same time his protector and friend, a more satisfactory relationship existed. James Tait had three rules for the discipline of negroes. The first was never to let a negro know he was to be whipped until you could get hold of him, the second was never to be in a hurry to punish after an offense had been committed, and the third was never to talk to a negro except on business.[44] If the master or overseer was a tyrant, negroes often became so terrified by his presence that they were incapable of doing their work properly. Usually, punishment was administered for neglect of work or for quarrels which arose among the negroes themselves. "Punished Joseph for not attending to the horses — Moses for picking 14 lbs of cotton and Adam for not attending to the cows. Whipped George severely for throwing his ax at Stephen. Whipped Tener for contrarriness [sic]."[45]

The practice of allowing negroes to have their own patches of cotton and vegetables, the products from which were to be sold to their masters, prevailed on many Alabama plantations. Contemporary experts in plantation economy condemned the practice as "nothing but evil, and that continually". Dr. N. B. Cloud contended that it opened a strong temptation to theft, because many negroes, who were unable to make good crops, proceeded to increase their yields by visits to the master's gin house or corncrib or those of his neighbors. Sometimes the evil began sooner than the gathering of the crop, for a privilege of working in the patch until nine at night was often construed to mean a visit to a neighbor's hen house

---

[44] *James A. Tait Plantation Memorandum Book*, 1852.

[45] *Burford Diary*, Jan. 2-15, 1850.

or hog pen.[46] The worst of the features, thought Dr. Cloud, was the fact that if one planter started the practice and it was not taken up by his neighbors, their slaves would sooner or later become dissatisfied. The prospect of from twenty-five to fifty dollars a year would seem large to the negro and cause him to exhaust his strength further after his day's labor in the field.[47]

Despite the arguments against this practice, Gould allowed his negroes on the "Hill of Howth" plantation to have their own patches of potatoes and cotton. In 1840 his cotton book recorded the following results of negro crops: "David 1140, Jordan 730, Phil 850, Bill 1000, Bob 947, Peter 1050, Augustus 500 and Squire 475."[48] These yields, aggregating thirteen bales, sold at Mobile for between eight and ten cents a pound and were marketed along with the master's crop of 126 bales.[49] No information is available to show whether or not the proceeds from these additional bales were ever turned over to their rightful owners or retained by Gould for purposes of safekeeping.

The soils of Alabama were admirably adapted for the production of the crop that placed Alabama first among the cotton-producing states of the country in 1850. It was generally felt that a loamy soil such as existed in the river bottoms and prairies was the best, for the ante-bellum planter was rather indifferent to any method of enriching and building up poor soil. With the exception of occasional use of barnyard manure, the only other fertilizer which was used was that made from rotten leaves. About the first of July several wagon loads of leaves and trash were spread on the ground and on top of this was placed a thin layer of dirt. A layer of leaves and a layer of dirt were then alternated until the pile

---

46 *American Cotton Planter*, New Series, II, 20.

47 *Ibid.*, II, 21.

48 *William P. Gould Papers*, Cotton Book, 1840.

49 *Ibid.*

reached a height of about three feet. In the middle of the pile a slight depression was usually made in order to hold the rain, which facilitated the rotting of the mass. By January the fertilizer was ready to be hauled to the field.[50]

Though the crop was not planted until March, the preparation of the ground usually began in February.[51] The fields were plowed deep and left in ridges about four feet apart. This was done in order to allow the rains to settle the soil before planting time. After the ground had been pulverized by a small harrow, the ridges were split open and the cotton planted in drills, which, when thinned, were from fifteen to twenty inches apart. Planting was always done by hand at the rate of about two bushels of seed to the acre, and the furrows were covered by means of a wooden scrape.[52] Frequently, late frosts in March killed the crop and planting had to be done over again, but in spite of the warnings of such experts as N. B. Cloud of Alabama and W. M. Philips of Mississippi, planters continued to gamble on the weather.[53] In 1822, Charles Tait wrote his son, James, that "the making of cotton is for me a poor business unless I can get to market sooner and sell it better. Perseverence [sic] is my doctrine in planting."[54]

When the cotton plants had attained a height of several inches the laborious process of thinning began, "chopping cotton" as this operation was called in Alabama. Two plants were usually left in each hill, and eventually one, about twelve to fourteen inches from the next plant. Cotton continued to be cultivated with the hoe and a plow

---

[50] *James A. Tait Plantation Memorandum Book.*

[51] *Ibid.*; *Cotton Book of the Thorn Hill Plantation,* March, 1853; J. A. Turner, *The Cotton Planter's Manual,* pp. 53-59; *Affleck's Southern Rural Almanac,* 1854, p. 12; *Harper's New Monthly Magazine,* VIII, 447, 462.

[52] *American Cotton Planter,* I, 202; *The Cultivator,* XI, 49.

[53] *Ibid.,* X, 94; *DeBow's Review,* Jan., 1857, p. 77.

[54] *Charles Tait Papers,* Jan. 15, 1822.

called the sweep, which threw a little light dirt into the place made by the drill. Weeds and grass had to be removed and the earth kept loose and well stirred around the stalk of the plant.[55] One negro was generally given from eight to ten acres of cotton to attend.[56] Cultivation of the entire crop continued at intervals of about twenty days until nearly picking time. In 1837, this notation was made in Charles Tait's plantation journal: "It took 14 days to plow over the whole crop the first time with 12 horses. 11 days to work it over with hoes the first time. The ground was very hard and stiff on the 'Over the River Place'."[57]

James Tait preferred to "list" his cotton with a scooter if the ground was hard and with a shovel plow if it had been softened by the rains. In the event that hard spring rains beat down the rows he built them back up again by running his turn plows. On stiff land this was particularly necessary in order to keep the seed well covered. After the cotton was thinned in the hills, sweeps had to be kept constantly running in order to keep the soil pulverized and the grass down. "Keep your plows a-going, a-going and your hoes a-going too", Tait often stated.[58]

The cotton-picking season usually commenced about the middle of August and ran intermittently until about the first of the year, though it was not unusual for the last scattering picking to take place as late as February. On the Gould plantation it was customary for the negro children to gather the last remains, which never amounted to more than a few bales.[59] The two essential requirements of picking were speed and a merchantable product

---

55 *American Cotton Planter*, I, 204.

56 *Ibid.*, I, 202.

57 *Charles Tait Journal*, 1837.

58 *James A. Tait Plantation Memorandum Book*, 1841.

59 *Cotton Book of the Thorn Hill Plantation*, Aug., 1852; *William P. Gould Papers*, Cotton Book, 1835-1836.

as free as possible from dirt and trash. Rapid picking prevented cotton from being spoiled in the field by rain or dew or by falling on the ground.[60]

For efficiency as well as convenience, each hand was supplied with two baskets and a bag usually made from a yard of cloth with an opening slightly lower in the front. As a rule, pickers worked four rows at a time and the basket was placed between the middle rows the distance from the ends necessary to insure the filling of the bag. Picking began at opposite ends of the rows and moved towards the basket, into which the bags could be emptied without loss of time, for no steps had to be retraced going to and from the basket. If the field was any distance from the gin house, the day's gathering was usually hauled in when the day was over, for no hand would do his best if he knew that when night came he would have to carry his basket a half mile or more.[61] Hands frequently were offered small prizes in the form of tobacco, handkerchiefs, molasses, and calico dresses as an inducement to put forth their best efforts, for during the picking season a large part of the planter's profit was at stake.[62]

Planters liked to boast about the ability of their expert cotton pickers and frequently made wagers with their neighbors. Contests were held in which the participants picked from sunrise to sunset under the supervision of a disinterested party.[63] In most cases, the ability of pickers was greatly exaggerated, for the average was about 100 to 120 pounds a day. The negro who could pick 500 pounds a day was a rare exception, though the cotton book of M. Harris Walker of Dallas County listed three negroes each of whom picked 300 pounds a day for six days.[64]

60 *The Southern Cultivator*, VII, 164.

61 *Ibid.*; *M. Harris Walker Diary*, Aug., 1854.

62 *The Southern Cultivator*, VII, 164.

63 *American Cotton Planter*, New Series, II, 328; *The North Alabamian*, Oct. 8, 1847.

64 August, 1843.

The most rapid progress in picking came in October and November after the frost had opened wide the bolls and caused a good many leaves to drop off the plant. Twenty-eight negroes on the "Hill of Howth" plantation picked 14,229 pounds in five days, which was an average of about five and one-half bales a day. The average amount per hand was about 100 pounds, but this included the work of several children who picked only from forty to eighty pounds a day.[65] L. M. Burford stated that his hands averaged from 120 to 150 pounds a day, but it was difficult to estimate accurately because of the large number of children who were always in the field during the picking season.[66] In the fall and winter of 1851, from five plantations in Wilcox County James Tait picked 752 bales of cotton averaging 500 pounds each. This was probably one of the largest single crops produced in Alabama before 1860.[67]

In 1856, William P. Gould became intensely interested in a mechanical cotton harvester invented by George A. Howe. Howe's agent, W. E. Cornwall of Mobile, wrote Gould and advised him to buy the machine, which he claimed could pick from 800 to 1,000 pounds of cotton a day at a first cost of twenty-five dollars a machine. The principle involved was that of an endless chain revolving over pulleys and propelled by a crank attached to the main pulley. Cotton was supposedly seized by the chain and deposited in a bag which the operator kept attached to the bottom of the machine. Gould wrote for further information and even asked for a demonstration on his plantation, but after many vague promises the agent finally refused.[68]

---

65 *William P. Gould Papers*, Cotton Book, Oct., 1855.

66 *L. M. Burford Papers.*

67 *James A. Tait Plantation Memorandum Book*, 1851.

68 W. E. Cornwall to William P. Gould, Feb., 1856, *William P. Gould Papers.*

The second most important crop on the Alabama plantation was corn. It was always planted in March and shared, along with cotton, the principal attention of the hoe and plow gangs. If planted in the early part of the month, the first process of weeding and thinning could be taken care of before the cotton was up. Again, in May, cotton had to be temporarily neglected in order to give the corn a thorough plowing and hoeing. Constant cultivation was necessary until the stalks began to tassel, for, as James Tait stated, "Good plowings are the main chance for making a crop of corn."[69] He also concluded that forty acres were necessary to raise 800 bushels, and his plantation memorandum book shows that in 1840 328 wagon loads of corn were raised on his two plantations known as the "Old Place" and the "Lower Place".[70] In a one-acre trial plot, Colonel Thornton, in 1853, succeeded in raising 120 bushels, which was an unbelievably high yield, considering that the soil was merely bottom land to which no fertilizer had been applied.[71]

Not only was corn the main article of diet for negroes on the plantation, but it was also essential as feed for mules, hogs, stock and poultry. After the crop was gathered in the fall, fodder was pulled, to be converted into an important part of the hay crop for winter feeding. Charles Tait found it convenient to estimate each year the anticipated consumption of corn on his estate. In the early thirties his yield appeared to be only twenty bushels to the acre. This, no doubt, was due to the fact that the best bottom lands were reserved for cotton. In 1832, when slaves numbered about seventy-two, his prospective needs were as follows:

> 785 for bread, at a peck of meal per week for each slave over ten years and half as much for young

[69] *James A. Tait Plantation Memorandum Book,* 1841.
[70] *Ibid.,* 1840.
[71] *Cotton Book of the Thorn Hill Plantation.*

children, plus the needs of the planter's family and the overseer
200 for stock hogs at fifty ears per day
200 for fattening hogs, 50 in number
375 for ten plow horses during five months
120 for four gin horses during four months
200 for 10 horses the balance of the year
300 for two horses and hogs at Dry Fork

2180 bushels as the total estimate[72]

The second important item in slave diet was pork, and hog killing on the plantation was hailed as a great ceremonial occasion. The cold days of December, January and February were the times usually chosen, but for the want of fresh meat it was not unusual for a hog or two to be killed even during the summer months. The number of hogs butchered was directly proportional to the number of negroes to be fed. During the year 1851 James Tait killed 340 hogs on three plantations,[73] while B. C. Yancey, with a much smaller estate in Cherokee County, killed fifty-eight in 1857 and sixty-three in 1860.[74] The plantation hogs, in many instances semi-wild, produced nothing like as many pounds of meat as the fine breeds of today. 175 to 200 pounds was about the average,[75] though Gould spoke of ten hogs killed in 1850 as averaging 244 pounds each.[76] Lard was always an essential by-product of the hog-killing season, and slave

[72] *Charles Tait Journal*, 1836. Professor Phillips estimates that Tait's corn consumption was 1,880 bushels, but he apparently failed to include the correct number of bushels used on the Dry Fork plantation, and he omitted the item of 200 bushels "for 10 horses the balance of the year". See *Life and Labor in the Old South*, p. 279.

[73] *James A. Tait Plantation Memorandum Book.*

[74] *B. C. Yancey Coosa River Plantation Account Book.*

[75] *Ibid.*; *M. Harris Walker Diary;* Susanna Clay to C. C. Clay, Sr., Dec. 6, 1826, *Clay MSS*; *Burford Diary*; *James A. Tait Plantation Memorandum Book.*

[76] *William P. Gould Papers*, Dec. 23, 1850.

women were often employed for several days rendering and drying, since this came during the off season in the fields.[77]

When the picking season was over or when the weather was too wet for work in the field, the planter found it necessary to turn his attention to odd jobs about the plantation. Tait found the winter months useful for breaking mules, moving negro houses (this was done every three years), repairing and constructing plantation buildings, cleaning out ditches, and clearing as much new ground as possible.[78] The constant repair of fences was not only a time-taking but a costly item, and this was even more true in the prairie counties where timber had to be hauled several miles.[79] During rainy days, women were put to work in the loom house and men to tanning leather or sorting cotton seed.[80] On the large plantations, fire wood was sometimes gathered in considerable quantities, for many planters felt that after a hard day's work the hands could not be expected to cut their own wood. Robert Withers wrote C. C. Clay, Sr., that his "overseer has been busily engaged all the week in cutting wood and expects to get enough to last all winter."[81]

Cheap lands, the frequent sale of insolvent estates, and the desire for increased profits tempted many planters to buy more land than they themselves could manage, which, no doubt, caused many to become interested in non-resident planting. The traveler, Lyell, noted that many plantations in Alabama were the property of North and South Carolinians, and that residents of Alabama owned lands in Mississippi.[82] The absentee planter had to depend entirely on the judgment of his overseer in

---

77 *Burford Diary,* Jan. 14, 1850.

78 *Charles Tait Journal.*

79 *American Cotton Planter,* I, 118-119.

80 *Affleck's Southern Rural Almanac,* 1854, p. 8.

81 *Clay MSS.,* Sept. 1, 1827.

82 Charles Lyell, *A Second Visit to the United States,* II, 71.

running the plantation, because if he resided in a distant state, his visits were never more frequent than once a year, and sometimes at even greater intervals. Thomas Glover of Orangeburg, South Carolina, who owned a plantation in Lowndes County, found it difficult to supervise his overseer by mail.[83] So involved did plantation affairs become in 1837 because of the drought, the cut worm and the indifference of the overseer that Glover sold his holding, for he was convinced the venture was costing him more than it was worth.[84]

---

83 Glover to McMichael, Sept. 21, 1837, Ulrich B. Phillips, *Plantation and Frontier Documents*, I, 185.

84 Same to same, *ibid.*, I, 184.

CHAPTER IV

PURCHASE AND CARE OF SLAVES

THE MANIA for buying slaves which seized Alabama planters is evidenced by the fact that the negro population of the State increased from 42,024 in 1820 to 342,884 in 1850, which was far in excess of the natural increase.[1] From the slave markets of Virginia and Maryland traders came to Alabama, in some instances displaying their wares in the larger centers of population such as Montgomery and Mobile, and in other instances moving from town to town.

During the same period the individual holding of James Tait grew from sixty-nine to 340, and there is evidence to show that the greater part of this gain came from purchases.[2] At the death of Charles Tait in 1835, his holdings, combined with those of his son, James, amounted to 115 slaves, of which eighty-six were adults and twenty-nine were children.[3] Shortly after the Taits moved to Alabama in 1818 they wished to increase their force of slaves by purchases on a low market. One dealer claimed he was getting $850 for likely field hands, but the elder Tait thought this price too high. He expected to make his crop, ship it to New York and then shop around for slaves at the best price. A negro trader by the name of Bohanneu told Tait that in Virginia negro men could be purchased for from $450 to $500 and women for less, and Charles advised James to go to Virginia, make his purchases and bring the negroes

---

[1] See Fourth and Seventh Censuses of the United States.

[2] *Manuscript Census Returns*, Wilcox County, 1820, 1850.

[3] *James A. Tait Plantation Memorandum Book*, 1835.

home.[4] By 1820 the Taits are known to have bought four negroes. A man named Len, his wife and their two children were purchased from Henry Hitchcock of Mobile for $1,250.[5] The next year other purchases were made, for James wrote his father as follows:

> That fellow Harry whom you bought of Tutt is fitified or subject to convulsions. If he is comitable [sic] I would make him take the negro back at once or sue him for the money. I presume that there is no doubt but that the money could be recovered. I will be glad if you will write to Tutt from Cahawba and tell him to write me.[6]

Among Charles Tait's early purchases was a negro of unusual intelligence named Hartford. While Tait was away on a business trip in the North he was kept informed about plantation affairs by this slave, who wrote a rather remarkable letter in an excellent hand.

> Mobile, Nov. 6th, 1826
>
> My dear Master,
>
> Your kind favor to me through Mr Caulborne has been duly recd and I now hasten to answer the same and inform you how your affairs are going here. I left the plantation on the 31st ultimo at which time they were coming on very well with the crops. Eighty bales had been picked and they think they will have eighty more. I think the cotton is much cleaner than it was last year (ample room). The corn crop is very good and I think they will have plenty for next year. I saw Master James the day I left who told me his family was well, and that he would soon take them to the plantation to stay a few weeks, and I am sorry to have to inform you that nine of the children have died at the plantation, mostly with the Hooping Cough. . . . There will be five births at the plantation, and among the number Nancy is to give birth to one, after a suspension of fourteen years. . . . Times have been so hard that I have made but little for myself, but I am in hopes

---

[4] Charles to James, Oct., 1819, *Charles Tait Papers.*

[5] *Ibid.*, April, 1821.

[6] *Ibid.*, March 26, 1822; also see Ulrich B. Phillips, *Life and Labor in the Old South*, p. 275.

> that I shall do better. . . . I have another son named after myself. . . . the respects of your effec. Svt. unto D[eath] in hopes to ever merit your esteem
>
> your most dutiful servant
>
> Hartford[7]

On the eve of the Panic of 1837 the Taits bought from White and Little, slave dealers, nineteen negroes for $16,750, payable January 1, 1838.[8] Though the record of purchases for the next four years is not available, it is evident that James more than doubled his holdings for his plantation memorandum book shows that he owned 275 negroes in 1841.[9]

In Madison County C. C. Clay, Sr., is known to have acquired between 1818 and 1830 twenty-three slaves, about half of whom were children, for a purchase price of $8,844. The price paid is slightly misleading, for a great part of this amount was paid in Tennessee bank notes, which fluctuated in value.[10]

Peyton Burford indicated his purchases in 1850 by the following notation in his diary:

> Nov. 11, 1850 — Paid $1500 for four negroes Mariah, — a woman 22 yrs and her boy William 6 yrs and Townsend age 3 and Henry age about one year, 'all of which negroes I warrent to be sound and healthy and the title good and complete in every respect whatever'. (Bought of S. D. Crum of Pleasant Hill, Dallas Co.) Dec. 17, 1850 — Received of Peyton Burford $1750 for the purchase of 2 negroes, a man 20 years old and his wife 18.[11]

The largest single purchase that the writer was able to discover in Alabama was made in April of 1860 and involved a payment of $32,000.

---

7 *Charles Tait Papers*, March 26, 1822. This letter, with certain variations, appears in *Life and Labor in the Old South*, p. 276.

8 *Charles Tait Journal*, 1837.

9 *James A. Tait Plantation Memorandum Book*, Jan. 1, 1841.

10 *Clay MSS.*, Bills of Sale, 1818-1830.

11 *Burford Diary.*

> Know all men by these presents that I, Curtis W. Jacobs of Worchester County Maryland have this day bargained and sold to Josiah Robins of Wilcox County Ala. the following named negroes as slaves for life — viz. Levin, Julia, Arnold, Hulda, Noah, Daniel, Maria, George, Sarah, Josephine, Soloman, Lench, Obed, Elisha, Roda, Fanny, Spencer, Aralanta, Lucy, Josiah, Lonize, Leach (child), Minnie, Kit, Clarissa, John, Frank, Thomas, Jacob, Charlotte and all of their increase there being several children whose names are not remembered. All of the said negroes I guarantee to the said Robins, his heirs and assignees forever, for the sum of thirty two thousand dollars ($32000) to be paid as stipulated in a note of hand I hold against him.[12]

The note was to bear 8% interest and mature on January 1, 1870.[13] However, with the coming of the war and emancipation, Robins believed that the whole transaction was null and void, but to his surprise Jacobs wrote in 1868 to get a final settlement of the amount still due him. Robins refused further payment and Jacobs brought suit, but the case was decided in favor of Robins.[14]

The usual method for rough estimation of the price of slaves was by the quoted figure of cotton per pound. For instance, if cotton was selling at ten cents, the price of a prime field hand was placed at $1,000, if at twelve cents, $1,200, etc.[15] In Alabama this general ratio remained fairly accurate except during two periods, namely, 1816-1820 and 1855-1860. In the earlier period, when cotton was bringing from sixteen to thirty cents a pound, the price of slaves averaged from $300 to $600 each,[16] and in the later period, with cotton around ten cents, the price of a prime field hand was

[12] *Josiah Robins Papers*, April 16, 1860.

[13] *Ibid.*

[14] Jacobs to Robins, Aug. 11, 1868, *Josiah Robins Papers*.

[15] Ulrich B. Phillips, *Plantation and Frontier Documents*, II, 74; *Florence Gazette*, Jan. 5, 1859.

[16] See *Hugh Rolston MSS.*, 1814-1816; *Deed Records*, Madison County, Book A; Greene County, Book A; *Clay MSS.*, Bills of Sale, 1818-1821; *Charles Tait Papers*, Oct. 18, 1819.

$1,600 and in some cases even higher.[17] Such prices after 1855 were about 30% higher than the average price of cotton warranted. However, there were exceptions to this rule. In October, 1860, a sale of thirty negroes took place in the town of Eutaw, in Greene County, and the highest price offered was $1,200 for a likely young field hand. Several women who the year before would have brought $1,400 were sold for less than $1,000. In commenting on the sale the editor of the *Alabama Beacon* stated that the prices offered were 30% to 50% lower than during the preceding year and attributed this slump to "an apprehension of a dissolution of the Union".[18]

Generally speaking, planters were reluctant to part with their negroes and did so only to satisfy their debts or for reasons of plantation discipline. Unruly negroes and those subject to running away were frequently sold to a passing slave dealer or to a neighbor who was in need of additional field hands. Advertisements of sheriff's sales, such as the following, often appeared in the newspapers: "To be sold to the highest bidder, for cash at the Court House door in the town of Moulton on Monday the 1st day of September next five negro slaves, Tom Eubank, Sheriff."[19]

If ready cash was not a necessity, planters would frequently take interest-bearing commercial paper. In December, 1853, H. L. Clay wrote C. C. Clay, Jr.: "I under stand [sic] that you are willing to give time — do not need the money just now and will be satisfied with

---

[17] See *Jacksonville Republican*, Oct. 7, 1857; *Alabama Beacon*, Dec. 18, 1857; *Daily Messenger*, Jan. 7, 1857; *Clarke County Democrat*, March 4, 1858, Jan. 20, 1859; *Hunt's Merchant's Magazine*, XXX, 500; *William P. Browne MSS.*, 1859; *Will of R. H. Gregg*, Wilcox County, 1860.

[18] *Alabama Beacon*, Oct. 26, 1860.

[19] *Moulton Democrat*, Aug. 28, 1856; also see *Eufaula Democrat*, Aug. 29, 1848; *Huntsville Democrat*, March 7, 1840; *Alabama Republican*, Oct. 20, 1820; *Tuskegee Republican*, Jan. 6, 1859.

bonds bearing interest from date . . . . Let me hear from you as soon as possible, because I wish to sell before the many other negroes advertised for sale are offered on the block."[20]

On certain occasions, planters from other states wrote their friends in Alabama in an effort to locate prospects for the sale of negroes. Such was the case when Thomas Flournoy, a planter-merchant of Augusta, Georgia, wrote Bolling Hall to find out whether Alabamians would buy his negroes. Flournoy stated that cotton was down to ten cents and at that price slavery was unprofitable in Georgia.[21]

The domestic slave trade thrived in Alabama as in other Southern states, and its two principal centers were Montgomery and Mobile. Though traffic in negroes was socially degrading, a native of Sumter County, Peter Stokes by name, was actively engaged in selling to Black Belt planters.[22] Stokes, who apparently operated out of New Orleans and Mobile, found that the fall was the best season for sales, for then the cotton crop was being gathered and planters had money or credit. During the year of 1846 he wrote from Gainesville, Alabama: "I will make from 21[00] to $2200 clear. I flatter myself that I can beat any living man selling negroes at this place. There has [sic] been 2 traders here for the last 4 or 5 weeks and I have sold some 25 and neither of them have [sic] sold a negro at all."[23] A month later, however, Stokes found the market dull at Mobile, for his letter home stated, "I can't sell at any price at all. I have got the same 3 I had when I wrote last if I don't sell in a few days I shall go back to Gainsville. They are not selling any over at Orleans either."[24]

---

20 *Clay MSS.*, Dec. 11, 1853.

21 *Bolling Hall MSS.*, 2nd Series, Jan. 3, 1823.

22 *Hatchett MSS.*, 1846.

23 Peter Stokes to D. R. Stokes, Feb. 8, 1846, *Hatchett MSS.*

24 Same to same, April 24, 1846, *ibid.*

In traveling about the State traders usually carried their slaves in wagons, pitched their tents and remained for two or three days in the small communities and for a week or more in the larger towns. The arrival of the trader was usually advertised well ahead of time, so that planters desirous of making purchases might be informed of his coming.[25]

Before the coming of the railroads in the fifties, Mobile was probably the most important slave-trading center of the State. Slaves could be brought by water from Maryland, Virginia and the Carolinas and delivered to purchasers up the rivers. It seemed only natural for planters who sold their cotton and bought their supplies in Mobile to purchase their negroes there. On the eve of the Civil War Mobile's trade had become exceedingly small and the recruits for its slave market came from estates of deceased or bankrupt planters in South Alabama or from small slave owners selling for profit or from necessity.[26]

In 1846, Lyell witnessed a slave auction in Montgomery in the vicinity of the Artesian Basin and seemed impressed by the fact that the next day a man was selling horses on the same spot.[27] From a rather meager beginning, and despite the competition of rival markets in Selma[28] and Columbus, Georgia, Montgomery grew to be one of the most thriving of southern markets. The city directory in 1860 listed four slave depots along with the same number of banks.[29] Apparently, one of the most active dealers in Montgomery during the fifties was Mason Harwell. From his advertisements in the newspapers one might infer that he was a general auctioneer,

---

25 B. F. Riley, *History of Conecuh County*, pp. 109-110.

26 This fact is rather clearly indicated in the Mobile newspapers from 1850-1860. Also see Frederic Bancroft, *Slave Trading in the Old South*, pp. 298-299.

27 Charles Lyell, *A Second Visit to the United States*, II, 42.

28 John Hardy, *Selma, Her Institutions and Her Men*, p. 16.

29 *Montgomery City Directory*, 1860.

for on his own account he bought and sold almost everything, including railroad stock. He also carried on various kinds of slave trading, including a hiring agency, and as a representative of the Knickerbocker Life Insurance Company of New York he was always ready to insure slaves at "reasonable rates".[30]

Despite the fact that William L. Yancey had long advocated a reopening of the slave trade,[31] Alabama's participation in foreign slaving activities was confined to a few illicit attempts of ships to slip into Mobile. In 1818 three vessels were captured and brought into Mobile for adjudication, their cargoes of 109 African negroes seized and the vessels confiscated. Some of the negroes were sold to a planter in Clarke County, while the remainder were disposed of locally.[32]

In 1859 occurred the last attempt to bring native Africans into Alabama. In that year three brothers, Jim, Tim and Burns Meaher, natives of Maine though at that time residing in Mobile, sent the schooner *Clotilde,* under the command of Captain Foster, to the coast of Africa, where a load of slaves was taken aboard. After an exciting homeward voyage, the schooner, under cover of darkness, slipped past the customs authorities at the entrance of the bay and made her way up to Spanish River, where the negroes were transferred to a tug and the *Clotilde* scuttled in Bayou Corne. The tug then proceeded up the Mobile River to its junction with the Alabama, where the negroes were unloaded at the plantation of John M. Dabney and hidden in the canebrake. When questioned by Federal authorities, the Meahers could account for their whereabouts, and therefore appeared not to have been engaged in the slave smuggling venture. It was on this alibi that Tim was admitted to bail pending in-

30 *Montgomery Advertiser,* 1854-1856; Bancroft, *Slave-Trading in the Old South,* p. 296.

31 *The Confederation,* June 8, 1860.

32 *Alabama Republican,* May 4, 1821.

vestigation, and while out on bond he completely outwitted the government authorities by disposing of the negroes, the proceeds from which were given to his supporters as their share in the enterprise.[33]

In caring for slaves, planters were in practically every case motivated by self-interest. However, it sometimes happened that through a feeling of genuine devotion slave owners stipulated in their wills that their negroes be freed. The will of Samuel Townsend of Madison County, recorded in September, 1856, is unusual in this respect. Townsend requested that thirty-seven of his slaves be emancipated at his death, and if the laws of the State of Alabama prevented this, that they be removed to a free state. The expenses of removal were to be paid by his executors out of a special trust fund, and if it were not practicable to remove all the negroes at once, the following named servants were to be moved first and emancipated: Westly, Thomas, Willis, Osburne, Joseph, Caroline, Elvoid, Parthena, Susanah and Milcha.[34]

As soon as the slaves were freed, he requested that they be moved to a permanent place of settlement, suggesting the colony of Liberia. When they were settled, a sum of $200 was to be given to each negro who was of age and an income from a general trust fund was to be divided equally among the minors. A special trust fund was also to be set up, the proceeds from which were to be applied from time to time toward the education and support of the negroes. Townsend's final request was, "I wish my trustees to keep the principal of my residuary trust fund so as to bring the greatest yield in a safe investment."[35]

---

33 For the best accounts see Emma Langdon Roche, *Historic Sketches of the South*, pp. 84-100; also Erwin Craighead, *Mobile: Fact and Tradition*, p. 358.

34 Will of Samuel Townsend, Sept., 1856, *Will Record*, Madison County, I.

35 *Ibid.*

To safeguard the health of negroes on the plantation both planters and overseers had to keep an ever constant vigil. Since it was impossible to carry out the plantation program without some degree of exposure to the hands at all seasons, it was highly important that their houses, food and clothing be carefully looked after.[36] Field hands were required to wear hats during the summer months, and comfortable beds, as well as sufficient cover for cold winter nights, were considered necessary for all the negroes. The picking season in particular was thought to be a period of strain on the constitution of the negro. Not only were the hours long, but constant bending over necessitated more frequent periods of rest. Usually, two or three hours were given at noon, and if breakfast was eaten in the field an intermission of from thirty minutes to an hour was allowed for the morning meal. In some instances owners granted a half day on Saturday to the women, in order for them to do their washing and clean up their cabins, for by the usual rule cabins were swept out each day and scoured once a week.[37] Overseers were repeatedly cautioned to see that both houses and quarters were kept as free as possible from filth.[38]

The journal of Charles Tait in 1836 indicates a very definite attempt to safeguard the health of the negroes on his plantation. The most important of his rules were as follows:

> 1. All negro houses should be moved every four years to prevent filth accumulations and cholera or dyptheria [sic].
> 2. Hot sun and fogs are thought to be injurious especially during the picking season.
> 3. Aug. and Sept. are the most sickly months. Negroes should never leave their houses til [sic] sunrise as long as cotton is green (it is thought that dew on the green plant is injurious to their health).
> 4. Cotton pickers must straighten up at noon and walk

36 *American Cotton Planter*, New Series, II, 293.

37 *DeBow's Review*, Oct., 1851, p. 371.

38 *American Cotton Planter*, New Series, II, 293-294.

to the house as relief. Extra rations of molasses during the picking season are thought to keep off fever.

5. An early breakfast is an aid to health especially during the sickly season.

6. In case of fever the most essential thing is to keep the skin moist and administer light nourishment.

7. In case of tooth ache dip cotton in tallow and stop up the cavity. Pain will soon cease and the old tooth will eventually rot out.[39]

One of the greatest contributory factors to disease among negroes was the condition of their houses. Small, low and filthy, such places were ideally suited for the breeding of germs. One subscriber of the *Alabama Beacon* in Greene County noted that a disease known as ship's fever had repeatedly occurred among his negroes, and he was convinced that it was caused by the condition of their cabins. He believed it extremely important that planters adopt some system under the operation of which their negro houses would be properly constructed and adequately ventilated, and living conditions among the negroes could be regulated.[40]

It was usually deemed necessary that negro houses be built two and a half or three feet from the ground, so that they could be conveniently swept underneath. Also, if there were any trash or filth deposited there, the elevation enabled the planter or overseer to see it. Some attempts were made to have houses neat and comfortable, but for a small family a cabin sixteen by eighteen feet was considered large enough.[41] In addition, it was important that negro dwellings be furnished with a suitable number of privies placed in such positions as not to contaminate the water supply of the quarters.[42] In order

---

[39] These entries were probably made by James Tait, since Charles had died in 1835.

[40] *Alabama Beacon*, Oct. 19, 1850.

[41] *American Cotton Planter*, New Series, IV, 80; *Southern Cultivator*, VIII, 163; *The Independent*, Jan. 16, 1858.

[42] *Southern Cultivator*, VIII, 163; *Farmers' Register*, V, 302.

to facilitate ventilation, planters preferred openings near the roof, which the negroes would not be so likely to chink up with rags, rather than sufficient window space which would be kept bolted up at all seasons.[43] Many cabins consisted of two rooms separated by an open space covered overhead, known as the "dog trot". The fact that the negro dogs usually slept in the open hallway probably gave rise to the name. On the Molett plantation in Dallas County, parts of the slave quarters are still standing, and the writer found that both double- and single-room cabins constructed of notched logs are still being occupied by descendants of the former plantation slaves.

In view of the unsatisfactory sanitary conditions created by negro houses, the Taits decided to replace their shanties with brick barracks. In the Journal of 1839 appear the following specifications:

> The house to be 300 ft long, the two walls to be 18 ft apart in the clear, and be 8 ft high. One chimney, but a double one, to each apartment of 30 ft in length . . . making each room 15 ft by 18 ft. Quantity of brick 130,000. Ten hands can make them in two months. Mr. P [McPhail] says he will attend to the making and burning of the brick and lay them or do all the brick work for $500.

The Journal does not state whether or not this work was ever carried to completion, and a visit to the "Old Place" failed to bear evidence that the job was ever done. Perhaps the depressed cotton market of the early forties was not without its effects.

Though the food of negroes varied somewhat from plantation to plantation, the fact still remains that their two chief staples of diet were pork and cornmeal. The naturalist, Gosse, recorded that each adult received one peck of cornmeal and three and a half pounds of meat

---

[43] *American Cotton Planter*, New Series, IV, 79.

per week, with an occasional addition of molasses and fruit when in season.[44] Another traveler found that on a large plantation in the Warrior-Tombigbee basin each man, woman and child was supplied each week with four pounds of bacon, one peck of cornmeal, one pint of molasses and three salted fish. In addition to a supply of fresh meat at various times during the year and vegetables in season, each man was allowed to raise on his own account pigs, poultry and grain.[45]

Practically all slave owners were of the opinion that meat was essential to the African constitution if a full day's work was to be done efficiently.[46] One Sumter County planter even went so far as to take into account the weight of bone in issuing meat to his negroes. "I give one pound a piece [to each negro] a day when they [his negroes] are eating the bony parts, four pounds a week when they are given the joints and three and a half pounds when weighing the middlings."[47] In addition, he gave as much bread as they wanted, all the milk on the place except the little used by himself and his wife, frequent issues of molasses, and he endeavored to raise an abundance of vegetables, all of which were given without deducting any meat or bread. Among the vegetables best suited for negro diet he found peas, Irish potatoes, yams, cabbage and turnips, all of which he considered nourishing and easy to raise. Each family received its weekly allowance and was allowed to prepare the food according to individual tastes.[48]

An owner in Macon County found that after a hearty meal of meat, vegetables and bread at noon, a supper of

---

44 *Letters from Alabama,* p. 253.

45 Charles Lanman, *Adventures in the Wilds of the United States,* II, 186.

46 William A. Smith, *Lectures on the Philosophy and Practice of Slavery,* pp. 296-297.

47 *Sumter County Whig,* June 24, 1851.

48 *Ibid.*

bread and molasses, with plenty of milk, was sufficient.[49] On the "Hill of Howth" plantation in Greene County, the overseer was instructed to issue rations of meat monthly and a peck of meal for each grown hand on a stated day of each week.[50] In calculating the corn consumption at Black's Bend in 1840, James Tait concluded that 797 bushels would be necessary to feed seven grown negroes, eleven children and four whites. This would allow over 2.7 pecks a week per person, which by comparison seems to be an unusually large amount.[51]

From three to four pounds of meat and a peck of meal, with vegetables and small amounts of molasses, were about the average rations for the adult negro. During the winter when vegetables were not to be had, an additional half peck of meal was sometimes added. Many of the authorities on plantation management were opposed to issuing a weekly allowance to each negro family and requiring them to do their own cooking. They believed that unless precaution was taken negroes would steal from each other and the master would be forced to give additional rations, since he could not expect efficient work from partially fed hands. If given the additional supply the negroes would soon learn how to impose on his kindness by intentionally trading off their meat and pretending it had been stolen. Another objection was, they stated, that the improvident would use up their supply of meat before the week was over and would be asking for additional rations.[52]

The fact that negroes had to do their own cooking encroached upon their free time, both at noon and at night. Many times the food was cooked in a hurry and was badly prepared, no matter how abundant the supply might have been. This was never the case if a planta-

---

49 *Macon Republican*, June 24, 1852.

50 *William P. Gould Papers*, Instructions to Overseers, 1837.

51 *Charles Tait Papers.*

52 *The Southern Cultivator*, VIII, 162.

tion cook was solely responsible for preparing meals.[53] However, it was, no doubt, true that the majority of planters in Alabama preferred to issue rations to individual families. Whether this came about from expediency or indifference one is not able to say.

A careful selection of clothing, especially during the winter months, was an important item in the care of negroes. Sometimes the supply on the plantation got low, as evidenced by the note of Carnothan, overseer of William P. Gould, who wrote, "For God sake Mr. Gould send the negro cloth up before the river gits [sic] too low to have it brought to Hayes Landing. It would be verry [sic] bad for me to stop the ploughs and send to Kirkpatricks for it. You have no idea how naked these niggers are."[54]

Ordinarily, furnishings for the plantation hands were bought along with the other supplies when the planter made his rather infrequent business trips. In Mobile, for instance, there were numerous dry goods houses which specialized in negro clothes. However, in 1836, James Tait had a standing order with R. G. Hazard of Peace Dale, Rhode Island, for four bales of thirty-six-inch cotton homespun or osnaburg to contain 3,000 yards at seventeen and one-half cents a yard. From other sources he bought 550 yards of woolen goods at thirty-three cents a yard.[55] Mrs. Tait supervised the making of this material into clothing. One woman was kept sewing all the time, as were four house servants when they were not engaged in their other work. Each man was supplied with two shirts and two pairs of breeches per year, and to the women two chemises and two frocks were given. The children were issued two cotton shirts and one woolen coat every fall. In the winter the men

---

[53] *Ibid.*; *American Cotton Planter*, IV, 118; *DeBow's Review*, XVII, 422-424.

[54] *William P. Gould Papers*, May, 1844.

[55] *Charles Tait Papers.*

were given a woolen jacket and breeches, and the women a jacket or wrapper. Both men and women were issued one pair of shoes each fall.[56]

Tait's estimate for the amount of cloth necessary for the garments was as follows: "For breeches 2½ yds of a yd wide homespun. For shirts 3 yds of the same. For chemises for the women 3 yds of yard wide homespun, for frocks 5 yards. For woolens for men 6 yds for jacket and breeches. For women 2½ yds for wrapper."[57]

In the late summer of 1842, John Erwin, of Greensboro, wrote Henry Watson and asked the latter to purchase 120 yards of material suitable for making coats and pantaloons. For this purpose Watson suggested "sugar wool jersey—a heavy grey twilled goods that I found better than anything I have ever purchased."[58] Erwin also wanted thirty-three yards of material suitable for Glasgow jeans, which garments he had found quite satisfactory the previous year. For the negro children he ordered forty yards of good substantial linsey, making 193 yards in all. In addition, the order was to consist of fifteen pairs of negro blankets of large size, five smaller pairs for children and twenty cotton handkerchiefs for negro women.[59]

A Sumter County planter found that negro blankets of better quality could be bought in New Orleans at a cheaper price than in Mobile.[60] However, a slave owner in Greensboro advocated buying negro brogan shoes in Mobile rather than in eastern markets.[61] Locally made shoes, he affirmed, were not only better sewed but the soles were reinforced with three rows of steel-headed

---

[56] *James A. Tait Plantation Memorandum Book*, 1839.
[57] *Charles Tait Journal*, 1836.
[58] Erwin to Watson, Aug. 15, 1842, *Watson MSS.*
[59] Same to same, *ibid.*
[60] *Sumter County Whig*, June 24, 1851.
[61] *Alabama Beacon*, Feb. 27, 1856.

nails, while the uppers, in addition to the usual stitching, were held together with copper rivets.[62]

Sickness and disease among negroes on the plantation not only meant loss from work but if allowed to become widespread were likely to result in serious losses of capital. Overseers were, from force of necessity, required to treat the commoner ills and were usually placed in charge of the plantation medicine chest. In most instances, planters required that an account be kept of negroes losing time because of illness, and the entries of Robert Bragg, overseer for James Thornton, are typical: "Andy lost two days with bile, George sick with fever three days, Charlotte with a risen [sic] on her foot four days, Little Sooky lost three days with bile, Joe lost three days with chills and fever, Surry lost three days with bowel complaint and lite [sic] fever."[63]

In 1856 William P. Gould noted in his diary:

> The negroes are nearly well again, but I shall be surprised if tomorrow passes without some new cases. Monday is always a bad day for sickness, especially if there is fruit of any kind to be found — blackberries are getting plentiful and wild plums are beginning to ripen. I know of no way to prevent the negroes from eating them except to lock them up on Sunday.[64]

In Alabama, September seemed to be the most sickly month of the year. Even as early as 1825, John Coffee, of Lauderdale County, observed that the doctor visited his plantation nine times during that month, while from January to October only twenty visits were made, and no visits were made from May to August.[65] Gould believed that dysentery was the most prevalent disease among plantation negroes,[66] while on the Walker estate,

---

62 *Ibid.*; also see *A. A. Coleman Plantation Memorandum Book*, 1854.

63 *Cotton Book of the Thorn Hill Plantation*, 1851.

64 *William P. Gould Papers*, Diary, June 15, 1856.

65 *John Coffee Papers.*

66 *William P. Gould Papers*, Diary, 1855.

in Dallas County, pneumonia seems to have occurred nearly every year for six years. In December, 1845, four deaths occurred among the negroes between the ages of twenty and thirty, and during the next spring two more.[67] River bottom plantations were also subject to visits of ague and fevers during the summer months. Malaria, however, could usually be fairly well controlled by the proper treatment with quinine and calomel.[68]

Generally speaking, suicide seems to have been uncommon among Alabama negroes. The only attempted case which the writer was able to find was that of a negro woman on the Burford plantation, who, after accidentally splitting her foot open with an axe, tried to kill herself.[69]

Bills for physicians' services occurred frequently among the items of plantation expense. In 1822, the negroes on the Coffee plantation were visited twenty times by Dr. William Bruce, who rendered a bill for $34.75.[70] Of this amount, five dollars was for an examination, $2.25 for bleeding, and the rest for visits and prescriptions. The year 1826 was even more sickly, for from January to December, Coffee spent $119.75 for medicines and doctors' bills.[71] Three years later, from July to November, Dr. Campbell made thirteen visits to the Coffee negroes at an average fee of two dollars, although some visits ran as high as five dollars.[72] Between the years of 1842 and 1847 James Tait paid a yearly average of $160 for prescriptions and doctors' services, and in 1846 his plantation memorandum book

---

67 *M. Harris Walker Papers*, Slave Book, 1840-1846.

68 G. H. Thomas to Jones Fuller, Feb. 18, 1854, *Fuller MSS.*; *The Southern Cultivator*, V, 107.

69 *Burford Diary*, Aug. 23, 1850.

70 *John Coffee Papers*, Dec. 31, 1822.

71 *Ibid.*, 1826.

72 *Ibid.*, 1829.

noted having employed "Drs. Matthewson, Erwin, Caldwald, Bacon and Nettles to attend my negroes."[73]

If the amount of the doctors' bills is any indication, sickness on the "Hill of Howth" plantation of William P. Gould reached frightful proportions. Two doctors were employed to attend fifty negroes and Gould paid $2,667.49 to one for visits from 1850 to 1856 and $348 to the other for services during the same period.[74] Pressing financial conditions prevented Gould from settling in full until 1856 and the above amounts were accumulated over the six-year period.[75]

It was not at all unusual for a country doctor to be a planter as well. Such was the case of Dr. Walker of Dallas County, who owned fifty slaves and several hundred acres of land. His income from professional visits to the neighboring estates amounted to $430 from November to July in 1842-1843, and $487 during the first six months of 1852.[76] The records are silent, however, as to the income from his planting interests.

Though the cost of medicine was usually included in the bill of the attending physician, general plantation remedies for the more common ailments were bought in considerable quantities. Tait's purchases in 1835 consisted of castor oil in gallon lots, epsom salts, spirits of turpentine, calomel, cream of tartar and a few less common medicines, which were bought in Mobile along with the other plantation supplies.[77] Gould found it necessary to supplement his yearly supply in 1860 by a purchase of $14.09 worth of medicines from Evans and Butler, local dealers at Eutaw.[78]

Not only from the standpoint of cost, but also because

---

[73] *Charles Tait Papers*, 1842-1847.

[74] *William P. Gould Papers*, 1850-1856.

[75] *Ibid.*, July 1, 1856; Jan. 1, 1857.

[76] *Medical Account Book*, 1842-1860.

[77] *Charles Tait Papers*.

[78] *William P. Gould Papers*.

of the distances required to travel, owners and overseers endeavored to become experts in administering home remedies. The plantation prescription book was sometimes regarded as a document of pride, even though many of the remedies were based on negro superstition. Cures for ague, bruises, burns, insect bites, bowel complaint, colic, croup, corns, dysentery and lock jaw were listed in the prescription book of Gould. His formula for cure of cancer is indeed curious. "Cancer—A sure cure—Boil the root of the leaved dock until it becomes quite soft and then as a poultice apply 3 or 4 times through the day. Also bathe with a decoction of dock, this simple remedy has produced an entire cure."[79]

Though planters of Alabama, as a whole, were not averse to the religious training of slaves, State laws dealing with the assembly of blacks and the licensing of negro preachers were strict. Preachers in all cases had to be members of recognized religious denominations or sects, and persons violating this law were dealt with severely.[80] The Alabama Baptist Association, at its annual meeting in 1850, charged the executive board with the duty of extending to planters and other slaveholders the offer of the association to cooperate in the work of supplying their slaves with the preaching of the Gospel. The resolution further stated that "intelligent masters with the light of experience before them will regard the communication of sound religious instruction [to their servants] as the truest economy and the most efficient police and as tending to the greatest utility, with regard to every interest involved."[81]

It was generally contended not only by the Baptists, but by other denominations as well, that the negro was the most neglected branch of society, since owners laid

---

79 *Ibid.*, Jan., 1842.

80 C. C. Clay, *A Digest of the Laws of the State of Alabama*, p. 541.

81 *Alabama Baptist Advocate*, Feb. 13, 1850.

most of their emphasis on physical rather than on moral and spiritual care. The State meeting of the Baptists was concluded with the recommendation that the association employ suitable ministers, who, with the concurrence of owners, would assemble slaves at any place on the plantation or at near-by churches designated by the owners in each community as suitable places for worship.[82]

The sentiments of the Alabama association were reëchoed by the *Southwestern Baptist,* which, in the fall of 1850, recommended that all churches make arrangements for their colored members to hold their own conferences, and that the ordinances be administered to them separately, but always under the control and supervision of the church to which they belonged.[83] However, despite the lament of the Baptists for the iniquities of the black brethren, little success seems to have been achieved in bringing the Gospel to the plantation through the work of traveling ministers. Certain groups within the State were opposed to such a movement,[84] and the plantation preacher, for the most part, continued to care for his flock.

William A. Smith, in his *Lectures on the Philosophy and Practice of Slavery,* stated that the "first and most important of all social relations is the marriage relation."[85] In general, the family life of the slave was almost entirely regulated by his master, and this regulation varied according to the moral ideas of the planter himself.[86] In most cases, Alabama owners required their men to take wives from among the women on the plantation rather than from those of some neighbor. Nevertheless, matches were frequently made by buying or selling

---

[82] *Ibid.*

[83] *Southwestern Baptist,* Dec. 25, 1850.

[84] *Tuskegee Republican,* Jan. 1, 1857.

[85] See Smith, p. 315.

[86] Frances L. Hunter, *Slave Society on the Southern Plantation,* p. 7.

one party or the other. J. B. Grace, overseer for Charles Tait, wrote him in April of 1835:

> I wish the three girls you purchased had been all grown, then they would have bin a wife apese for Harris and King and Nathan. Harris has June for a wife and Nathan has Edey — but King and Nathan had some difficulty hoo would have Edey. I promised King that I would indeavor to git you to buy a nuther woman so he might have a wife at home. He is envious that you git a nuther woman as he has quit the wife he had . . . Edey for a wife. I am willing to take Violet at the same price you paid for her and pay you her expenses in bringing her from Mobile — or any other expence you wish as she is too young [to make] any of your men a wife. You have small girls enough for your young boys — and then you can buy a woman in her [Violet's] place.[87]

A writer in *DeBow's Review* felt that marrying among hands on the plantation should be encouraged as much as possible, but that intermarrying with slaves belonging to other estates should not be prohibited, though the results were usually not good and should be discouraged as much as possible.[88]

As negroes increased in age their tasks were year by year lessened, until, finally, infirmity and old age made them no longer useful in the routine of the plantation. When this time arrived they were usually assigned to such tasks as minding the little negroes and cleaning up trash from the planter's yard and premises, or were allowed to while away the day according to their own inclinations. In very few instances were they looked upon in a different light from the other members of the plantation force as far as food, clothing, housing and medical care were concerned, and in this respect they probably had the most complete security of all classes in society.[89]

---

[87] *Charles Tait Papers.*

[88] *DeBow's Review,* XVII, 425-426.

[89] *Clarke County Democrat,* Dec. 3, 1857.

CHAPTER V

SLAVERY AND THE LAW

THOUGH Toulmin's *Digest*, published in 1819, was the basis of the slave code in Alabama, from time to time public opinion demanded more stringent laws in regard to slaves. This was particularly true after 1831, when the news of the Southampton Insurrection reached Alabama. Newspapers immediately began an active campaign for a more effective patrol system and called on the legislature for new laws to control more adequately the conduct of slaves.[1] Again, in the fifties, there was a general tightening of the administration of slave law. The activities of the abolitionists in the North and the increasing fear of the free negro brought about a renewed emphasis upon controlling the negro on the part of both the legislature and the law enforcement machinery of Alabama.[2]

Even during the Territorial period there were numerous restrictions on the activities of slaves. "No slave is allowed to keep or carry any gun, powder, shot, club or other weapon whatsoever . . . . except tools given to work with, or that he is ordered by his master, mistress or overseer to carry. . . ."[3] Every offender against this

---

[1] *Greene County Gazette*, Jan. 5, 1832.

[2] See *Acts of the General Assembly of the State of Alabama*, 1850-1856.

[3] *Mississippi Territorial Statutes*, March 6, 1805, Section 4; Harry Toulmin, *A Digest of Laws of the State of Alabama*, p. 627; Henry Hitchcock, *The Alabama Justice of the Peace*, p. 422; *Southern Advocate*, Jan. 2, 1829; *Mobile Commercial Register*, Jan. 14, 1829.

law was to be sentenced by a justice of the peace to receive not exceeding thirty-nine lashes on his bare back.[4]

Trading with slaves was also prohibited. No person was allowed to buy, sell or receive from a slave any commodity whatsoever without the consent of the master, owner or overseer, and any person who did so was forced to pay the owner of such slave four times the purchase price of the article received, in addition to court costs.[5] The slave committing such an offense was to receive, by the order of a justice of the peace, ten lashes from his master.[6]

Barkeepers who employed slaves to aid in drawing or vending liquor were subject to a fine of from twenty-five to fifty dollars,[7] and this same fine was also applicable to anyone who sold spirituous liquors to a slave without the written permission of the owner or overseer, specifying the quantity previously sold to that slave.[8]

The law was specific in preventing slaves from owning or keeping such domestic animals as dogs, horses or mules, and hogs.[9] An exception was sometimes made in the case of dogs and swine, provided it met the approval of the several owners in the neighborhood, but never were the hogs of slaves permitted to run at large.[10]

---

[4] Toulmin's *Digest*, p. 627.

[5] *Ibid.*, p. 629.

[6] *Ibid.*; John G. Aikin, *A Digest of the Laws of the State of Alabama*, p. 396. In 1845 a law was passed prohibiting any owner, trader or agent on any water craft on the navigable streams of the State from bartering or exchanging dry goods, groceries or spirituous liquors with any slave under penalty of $500 fine and twelve months in prison. *Acts of the General Assembly*, 1845, p. 154.

[7] *Acts of the General Assembly*, 1849-1850, p. 49.

[8] *Ibid.*; C. C. Clay, *A Digest of the Laws of the State of Alabama*, p. 543; *Worrell* vs. *The State*, 12 Ala. 732.

[9] Toulmin's *Digest*, p. 631; Clay's *Digest*, p. 542; Aikin's *Digest*, p. 394.

[10] Toulmin's *Digest*, p. 631; Aikin's *Digest*, p. 394.

Masters who allowed their slaves to hire on their own time or trade as freemen were subject to a fine of fifty dollars.[11] In like manner any master, mistress or overseer who knowingly allowed any slave, not his own, to remain on his premises for more than four hours without the leave of the owner or overseer of such slave, had to forfeit and pay ten dollars for every such offense.[12] "No slave shall go from the tenement of his master . . . . without a pass or some letter or token, whereby it may appear that he is proceeding by authority from his master, employer, or overseer . . . ."[13] Any slave violating this law might be apprehended and carried before a justice of the peace, to be by his order punished with not more than twenty stripes.[14]

As a direct result of the Southampton Insurrection, the General Assembly of the State of Alabama passed several acts in 1832 to strengthen the slave code. In that year a law was passed prohibiting any person from endeavoring to educate slaves, violators to be subject to a fine of not less than $250 nor more than $500.[15]

During the same session other acts of similar nature were passed. The death penalty was prescribed for anyone who should distribute, circulate or publish any "seditious papers, pamphlets, or writing, tending to produce conspiracy or insurrection or rebellion among the slaves or colored population . . . ."[16] The same act made it unlawful for free persons of color to associate with slaves in the kitchen, outhouses or negro quarters without written permission of the master, and those so offending were

---

11 Toulmin's *Digest*, p. 629; *Starr* vs. *The State*, 25 Ala. 49; *The State* vs. *Gaus*, 9 Porter 638.

12 Aikin's *Digest*, p. 392; *The State* vs. *Newman and Levy*, 7 Ala. 69.

13 Aikin's *Digest*, p. 391.

14 Clay's *Digest*, p. 539.

15 *Acts of the General Assembly*, 1832, p. 16.

16 *Ibid.*, p. 17.

to receive fifteen lashes for the first offense and thirty-nine lashes for every one thereafter.[17]

In 1832 it became unlawful for more than five male slaves, either with or without passes, to assemble together at any place off the proper plantation to which they belonged, and it was the duty of all civil and military officers to disperse such assemblies. Each of the slaves so assembled was liable to receive any number of lashes not exceeding ten.[18] Any officer of the law who had knowledge or information of such unlawful meetings of slaves and failed to suppress them was to be subject to a fine of ten dollars by any justice of the peace.[19] By the same act, no slave or free negro was allowed to exhort, harangue or preach to any assembly of slaves unless he was licensed by some regular body of professing Christians in the immediate neighborhood to whose society or church such negro properly belonged.[20] For any violation of this law the offender, in the presence of five slaveholders, was to be punished with thirty-nine lashes for the first offense and fifty for every subsequent one.[21]

A more adequate control of slaves immigrating into the State was brought about by a law in 1832 which required that within thirty days after arrival a statement had to be filed with the clerk of the county court designating the name, sex, and as nearly as possible the age, size and color of all slaves brought in. This was to be accompanied by an oath that the slaves were brought in for personal use and not for the purpose of sale, hire or

---

[17] *Ibid.*; *Alabama Beacon*, June 10, 1848.

[18] *Acts of the General Assembly*, 1832, Sec. 19, p. 18. The provisions of this act did not apply to assemblages of slaves attended by master, owner or overseer.

[19] Clay's *Digest*, p. 541.

[20] *Acts of the General Assembly*, 1832, p. 18.

[21] Henry C. Semple, *The Code of Alabama*, p. 239; "Nothing in this Act shall be so construed as to prevent free persons of color and slaves from attending places of public worship held by white persons." Clay's *Digest*, p. 544.

traffic. For every slave imported in violation of the law a fine of $200 was placed on the owner.[22] Because of obvious difficulties in enforcement, this law was repealed within a short time. In 1846 the General Assembly again took action on the introduction of slaves into the State. A special committee appointed to investigate the problem stated that the slave population was becoming too large for the interest and security of the citizens of the State. The committee reported further that non-resident planters sent their slaves into Alabama and withdrew the proceeds of their labor to be expended in other states. Fear was expressed that, as public opinion was setting itself against the continuation of slavery in the Upper South, Alabama might be overrun with the slaves of absentee planters.[23]

The Alabama legislature, in 1846, imposed a higher tax on slaves belonging to non-residents than was placed on those owned by residents. Much complaint arose from absentee planters who had slave property in Alabama. Judge Chilton of the Alabama Supreme Court delivered a unanimous opinion of the court which pronounced the act constitutional. Chief Justice Collier concurred and delivered a distinct opinion, from which the following is an extract:

> Slaves, it must be conceded, are a unique and peculiar description of property and that it is competent for the legislature to enact regulations of internal police in respect to them—which may discriminate between the resident and non-resident master. These measures of police may be so framed as to subject the non-resident to heavier pecuniary burdens and perhaps it is within the competency of legislation to prohibit all persons but immigrants from bringing and settling their slaves within this state.[24]

Rigid regulations had to be complied with in regard to transporting slaves by water. Before a vessel could be

---

[22] *Acts of the General Assembly*, 1832, pp. 13-14.

[23] *State Journal and Flag*, Jan. 23, 1846.

[24] *Independent Monitor*, Dec. 21, 1848.

cleared, a manifest of slaves issued by the port inspector had to contain the following information: name of slave, sex, age, stature, class (light or dark), shippers or owners, and residence. Any violation of this law placed a penalty on both the port inspector and the captain of the departing vessel.[25]

The patrol system was a special police for slaves for the purpose of preserving peace and order. By an act passed in 1819 the basic patrol law of the State was established, which specified that "every male owner of slaves, eligible to perform military duty, is eligible to perform patrol duty".[26] It became the duty of every captain of infantry to make a complete list of all male persons within the confines of his company eligible for patrol duty, and by rotation to appoint a list of patrol detachments, with the proper officer or leader of each. A detachment was to consist of no fewer than three nor more than five men, who were to perform their duties once a week for two months.[27]

It was the duty of each patrol to visit negro quarters and all places suspected of entertaining unlawful assemblies of slaves. Any negro found without a pass from his owner or overseer was to be given fifteen lashes.[28] In event the disorderly person was free, he was to be taken before the nearest justice of the peace and dealt with according to law.[29] If slaves were caught running away they were to be brought to the nearest justice of the

---

25 *Manifest of Slaves*, Jan., 1856, U. S. Customs House, Mobile, Alabama.

26 *Acts of the General Assembly*, 1819, p. 35; Toulmin's *Digest*, p. 639. Persons appointed for patrol duty were at liberty to send a substitute to perform in their stead. Toulmin's *Digest*, p. 639.

27 *Acts of the General Assembly*, 1819, p. 36; John V. Denson, *Slavery Laws in Alabama* (Alabama Polytechnic Institute Historical Studies, Third Series, 1908), p. 11; Toulmin's *Digest*, pp. 635-639.

28 *Acts of the General Assembly*, 1819, p. 36.

29 *Ibid.*, p. 37; Denson, *Slavery Laws in Alabama*, p. 12.

peace, where the owner, upon the payment of ten dollars, could take up his slave.[30] When information was received by a patrol of any persons harboring or concealing negroes, the patrol was supposed to go immediately in search of the negroes. If a conviction for such an offense occurred, the person offending was fined not less than $700 and one year in prison.[31]

Any member of the patrol who failed to report for duty could be court-martialed and fined not more than ten dollars.[32] In the event a patrolman disobeyed the orders of his leader or otherwise behaved in a disorderly manner, he was to be fined by the judgment of a justice of the peace a sum not exceeding ten dollars. This money was to accrue to the leader and other patrolmen of the detachment in the proportion provided for other fines. Furthermore, such a disobedient or disorderly patrolman was liable for the damages which happened to any individual in consequence of his disorderly conduct, payment for such damages to be recovered before any court having jurisdiction.[33]

On November 5, 1831, the editor of the *Southern Advocate* at Huntsville attacked the existing system of patrol laws as follows:

> The present patrol laws are very lame. On the first day of our Circuit Court I was surprised to hear the remarks made, I should say threats against the patrollers . . . . If the patrollers are not clothed with powers to correct negroes for insulting them, when in discharge of their duty, they had as well remain at home where they will be out of reach of indictments, and those gentlemen who own large numbers of negroes can enjoy themselves, whilst their negroes are eating the poor man's hogs and chickens.

---

30 *Acts of the General Assembly*, 1819, p. 37.

31 Toulmin's *Digest*, p. 641; *Acts of the General Assembly*, 1819, p. 38.

32 *Ibid.*, p. 37.

33 Toulmin's *Digest*, p. 637.

Other newspapers took up the issue of reforming the patrol laws. Another Huntsville editor stated:

> We must have a patrol, a regular efficient patrol at all hazards and upon any terms. It is madness to remain longer without it. It is our opinion that it will never be had, until militia organization, discipline and duty are vastly better enforced than at present, until reasonable compensation be given to those who do patrol duty at public cost.[34]

In 1841, at the request of Baldwin County, a special local patrol law was enacted. Slaveholders in each beat were divided into four groups, each under the control of a deputy-marshal, to do patrol duty. Each person liable to military duty was to serve one year out of every three and, for omission of duty, a fine of fifty dollars was to be assessed.[35] The town of Wetumpka, in 1842, organized a municipal patrol under the command of the town marshal. All slaveholders residing within the limits of the town were eligible for duty and organized according to wards.[36]

The laws of Alabama relating to runaway slaves were less harsh than those of some of the other Southern states and did not overstep the bounds of conservative measures for the recovery of property.[37] Numerous advertisements of runaways appeared in Alabama newspapers, some in the nature of advertisements by masters for the recovery of their property and others by justices of the peace endeavoring to locate the owners of negroes who had been apprehended and committed to jail. Typical of the second class is the following:

> Committed to the jail in Greene County as a runaway on the 7th inst. by W. W. Moffit a Justice of the Peace of the said County a negro man slave who calls his name

---

[34] *Huntsville Democrat*, Sept. 9, 1835.

[35] *Acts of the General Assembly*, 1841, p. 74.

[36] *Wetumpka Argus*, March 9, 1842.

[37] Denson, *Slavery Laws in Alabama*, p. 7.

> Frank and says he belongs to William Lewis who lives near Huntsville, Alabama. His owner is requested to come forward and prove his property, and pay the charges or he will be dealt with as the law directs.[38]

Any slave who was absent from his place of residence or owner's service without leave was considered a runaway. The owner or overseer was required to hand in to the nearest justice of the peace the name and the description of such slave within ten days after the fugitive absconded. Failure to comply with this law meant that the owner had to forfeit a dollar for every day that the information was withheld.[39] Runaway slaves could be lawfully apprehended by any person and carried before the nearest justice of the peace. When this was done, the justice required the slave to state his name and the name and place of residence of his owner, who was then notified by letter at his post office. If that was unknown, the communication was addressed to the county seat of the county in which he resided.[40] The owner, upon paying six dollars to the person who apprehended the runaway, plus other reasonable charges, could claim his property.[41]

In case it appeared impossible to find the name and residence of the owner, the sheriff was to advertise without delay in a newspaper published in the county an accurate description and other facts important to the fugitive's identification. Such an advertisement had to appear once a week for six weeks if he was not reclaimed sooner by the owner.[42] Upon payment by the master of the advertising costs and all other expenses attending the imprisonment, the judge of probate, by order in writing, was to direct the jailer, if the slave had not been sold, to

---

[38] *Greene County Gazette*, May 12, 1830.

[39] Toulmin's *Digest*, p. 630. A slave who was found eight miles from his master's plantation without a pass, or one who had lain out more than two days without leave was considered a runaway.

[40] *Acts of the General Assembly*, 1849-1850, p. 52.

[41] Toulmin's *Digest*, p. 630; Clay's *Digest*, p. 541.

[42] Semple's *Code*, p. 240.

deliver him to the applicant. If he had been sold, the order had to be directed to the county treasurer to turn over to the owner the proceeds of the sale.[43] If a fugitive was committed to jail and was not claimed by his owner within six months after the publication of his commitment, he was to be sold by the sheriff at public auction, the proceeds going to the county. If a runaway slave was sold out of jail and the proceeds paid to the county, the owner had the right any time within five years to demand and recover the sum received, upon the proper proof of his ownership.[44]

Many times, runaway slaves hid out and lurked in the woods and swamps, killing hogs and committing other injuries to the inhabitants of the country. In such cases, any justice of the peace of the county where the slave was supposed to lurk could direct the provost of the patrols to take such power as he deemed necessary for the effectual apprehension and commitment of such out-lying slave to the jail of his county for further trial. For every such slave apprehended and committed to jail the provost was entitled to a reward of thirty dollars, which had to be paid by the owner as a tax for the current year.[45]

In the event a slave committed manslaughter on the body of another slave or free person of color, he was to receive no fewer than thirty-nine lashes nor more than 100 at the discretion of the jury trying the case. Moreover, such slave might be branded on the forehead with the letter "M".[46] In the case of *The State* vs. *Peter*

---

[43] *Ibid.* "Children born in this state of a negro woman who is a fugitive slave are not fugitive slaves, or slaves which have escaped from service in another state, within the meaning of the Federal Constitution and the Act of Congress of 1793." J. W. Shepherd, *Digest of Alabama Reports*, p. 743.

[44] Clay's *Digest*, p. 542; Toulmin's *Digest*, pp. 634-635; *Acts of the General Assembly*, 1848, p. 139.

[45] Aikin's *Digest*, p. 394.

[46] *Acts of the General Assembly*, 1826, p. 42; *Huntsville Democrat*, Jan. 19, 1827.

(1827),[47] the prisoner was indicted for the murder of another slave and found guilty of manslaughter. The court passed judgment that he should be whipped and branded. In a capital case, two-thirds of the jury trying the slave were required to be slaveholders. For this type of case the master was considered a competent witness for his slave.[48]

In 1847, a case came before the Supreme Court of Alabama involving a slave who had resisted by violence an order to work, on the pretense that he was ill. The court held that the master or overseer was the proper person to judge whether or not a slave was too sick to be able to work. The latter could not, therefore, resist the order of the former to go to work for this reason.[49]

The laws of Alabama relating to the trial of slaves were explicit, and on the surface seemed to guarantee justice to the criminal. Before 1812, there were no juries for slave trials. By the act of 1807 it was provided that justices of the quorum of every county were to be justices of *oyer* and *terminer* in the trial of slaves charged with treason, felony or other crimes or misdemeanors. There were to be three justices of the quorum, who, in conjunction with two justices of the peace, should conduct the trial, which was to be without a jury. To convict a slave, all justices sitting had to agree in the verdict, and the prisoner had to have counsel to defend him. No justice who had an interest in the slave was allowed to sit in the trial. Legal evidence was defined to be "the confession of the offender, the oath of one or more creditable witnesses, or such testimony of negroes or mulattoes with pregnant circumstances, as to them shall be convincing".[50]

This act remained in force until 1812, when a superseding law was enacted which made juries necessary in all

---

[47] 1 Stewart 38.

[48] *Spence, A Slave,* vs. *The State,* 17 Ala. 192.

[49] *The State* vs. *Abram, A Slave,* 10 Ala. 928.

[50] Toulmin's *Digest* (1807), p. 328.

important slave trials. That law provided that for the trial of slaves charged with treason, felony or misdemeanors a competent court should be any three justices of the quorum with one more justice of the peace, excepting the justice who committed the prisoner to jail,[51] and a jury of men, no one of whom was the master of the slave or related to the master or to the prosecutor, and the court was to proceed without indictment.[52] This law was in effect only two years, when another act provided that a slave charged with an offense, not capital, might be tried by a justice of the peace on warrant and might be punished with not exceeding 100 lashes, provided two slaveholders concurred with him in the propriety of the sentence.[53]

The constitution of 1819 specifically states that "in the prosecution of slaves for crimes of a higher grade than petit larceny the General Assembly shall have no power to deprive them of an impartial trial by a petit jury".[54] For offenses below petit larceny the trial was to be as it was under the Territorial Act, except that not more than fifty lashes were to be administered by any justice.[55] Whenever any slave was brought before a county court for the commission of any offense of a higher grade than petit larceny, such slave was immediately committed to jail. The sheriff of the county was then directed to summon the justices of the county court and a jury of twelve "good and lawful men".[56] Neither the master nor

---

[51] The provision prohibiting the committing justice from sitting on the trial was to insure impartial judges. A justice might commit a slave on any charge, and it was to prevent such justice from sitting as a judge in the trial of a case in which he might be personally interested and thus have beforehand a fixed opinion in the case.

[52] *Digest of Mississippi Territorial Statutes* (1816), p. 192.

[53] *Ibid.*, p. 199.

[54] *Constitution of 1819*, Article 6, Sec. 2.

[55] *Acts of the General Assembly*, 1819, p. 88.

[56] *Ibid.*, pp. 88-89.

anyone related to him could serve on the jury and the trial was to proceed without presentment or indictment. No slave could be condemned unless he was found guilty by a jury after having been represented by adequate counsel.[57] The fee of the defense counsel, amounting to ten dollars, was to be paid by the owner of the slave. In case the judgment of death was passed upon the offender, it was required that at least thirty days should elapse between the passing of the judgment and the day of the execution, except in cases of conspiracy, insurrection or rebellion.[58]

In capital cases, when the jury returned the verdict of guilty, the same sworn jury was compelled by law to assess the value of the slave and the verdict was entered in the records of the court. The owner of the slave was to file with the comptroller of public accounts a transcript of the record of the court and was entitled to receive a warrant on the treasury for one-half the amount assessed by the jury, to be paid out of funds provided for the purpose.[59]

To raise this fund it was the duty of the assessors to levy a tax of one percent on all negroes under ten years old and two percent on all negroes over ten and under sixty, which was the duty of the tax collector to collect and return in the same manner as any other State tax.[60] The owner was required to employ good and sufficient counsel and if he did not, it became the duty of the judge to appoint such counsel to defend the slave. In cases involving rebellion, no fee was paid the owner.[61] Some

---

[57] *Ibid.*, p. 89.

[58] *Ibid.*

[59] *Acts of the General Assembly*, 1824, pp. 41-42.

[60] *Ibid.*

[61] *Ibid.* "If any slave escaping from any other state or territory shall be convicted and executed for crime committed in Alabama, the owner of said slave shall not be entitled to receive compensation." *Acts of the General Assembly*, 1846, p. 27.

minor changes were made in the trial procedure of capital offenses in 1832, and in 1841 the law was as follows:

> The trial of all slaves for capital offenses shall be by the circuit court of the proper county, in the mode now provided by law for the trial of white persons; but on such trial the slave shall be allowed but twelve peremptory challenges, and the State but four, and at least two-thirds of the jury shall be slave holders; but if there be an actual or threatened rebellion, of the slaves of this State, it shall be the duty of any justice of the peace before whom any slave or free person is brought, charged with a capital offense, if there be probable grounds for believing him or her guilty of the offense charged, to commit him or her to the county jail and appoint a time for the trial of the offense, not more than fifteen days from the commitment . . . .[62]

It might also become the duty of the justice to give notice of the time of such trial to the judge of the county court and some other justice of the peace. If there was no judge of the county court, one other justice was to be summoned in his stead and the majority of the persons thus summoned were to constitute a court for the trial of the offender.[63]

The justice making the commitment was required to have the sheriff summon a jury to appear at the time appointed for the trial of the accused, two-thirds of whom were to be slave-holders, and if the panel were exhausted by challenges or otherwise, the deficiency was to be made up from bystanders, but no one having an interest in the slave was to be considered a competent juror.[64]

In trials involving cases of insurrection, the only form of indictment necessary was a brief written statement

---

62 *Acts of the General Assembly*, 1840-1841, p. 189.

63 *Ibid.* "All slaves shall be bailable before conviction except for capital offenses, or when the proof is evident or the presumption great." *Acts of the General Assembly*, 1849-1850, p. 50.

64 See *ibid.*, Sec. 12, p. 189; Clay's *Digest*, p. 473.

of the nature of the crime, signed by the prosecuting attorney.[65] However, in regular trials the indictment had to be the same as in the prosecution of a white man; that is, it must have been returned by a legally constituted grand jury, and a copy of the indictment must have been delivered to the prisoner at least two days before his trial, and his counsel, which was to be provided by the court, if necessary, was to have a list of the names of the jurors summoned.[66]

If a slave was found guilty in a capital case, the court was to pronounce the sentence and cause execution to be done in the same manner as that prescribed for white persons, except that on conviction for conspiracy, insurrection or rebellion, execution might be held immediately.[67] Trials of capital cases as described above were to be held in the courthouse of the county, and the clerk was to keep a record of the proceedings in the minutes of the court.[68]

In 1841, trials of slaves charged with petit larceny or lesser offenses were conducted according to Territorial law, that is,

> For the offense of petit larceny, or any offense of a lesser grade, any slave may be tried by any justice of the peace on warrant, and may be sentenced to receive any number of stripes not exceeding fifty, which sentence shall be executed by the constable; but no justice of the peace shall be authorized to inflict more than thirty-nine lashes, unless he associates with him at least two respectable slaveholders, who will concur with him in the propriety of the sentence.[69]

---

65 *Acts of the General Assembly*, 1840-1841, p. 189; *Penal Code of the State of Alabama*, 1841, Ch. XV.

66 *Alabama Code of 1852*, Sec. 3319 and Sec. 3576.

67 *Acts of the General Assembly*, 1841-1842, Sec. 7.

68 *Ibid.*, Sec. 14.

69 *Ibid.*, Sec. 17; Clay's *Digest*, p. 474; *Penal Code of the State of Alabama*, 1841, Ch. XV.

All offenses higher than petit larceny, except capital cases, which, if committed by a white person or a free negro would be punished by imprisonment in the penitentiary, were to be tried by the judge of the county court and two justices of the peace, with a jury, and the jury was to direct the punishment, which was not to exceed thirty-nine lashes or branding in one hand, or both.[70] When a slave was convicted of an offense not capital, he was liable to be sold by the sheriff ten days after conviction to satisfy the costs of prosecution unless such costs were paid by the owner.[71]

No cruel or unusual punishment was to be inflicted on any slave and any person guilty of doing so was subject to a fine of $200.[72] If a person without just cause (to be determined by the jury trying the case) committed assault and battery on the body of a slave, such person, if not a member of a patrol, owner, overseer, or other agent, was liable to the same kind of punishment as if the offense had been committed against a free white person.[73]

The penal code of Alabama, 1841, states that

> If any person shall with malice or aforethought cause the death of a slave by cruel, barbarous or inhuman whipping, or beating or by any cruel and inhuman treatment, or by the use of any instrument in its nature calculated to produce death, such killing shall be deemed murder in the first degree.[74]

---

[70] *Acts of the General Assembly*, 1840-1841, p. 190.

[71] *Ibid.*, 1843, p. 121. The right of appeal in capital cases was not denied slaves. There was no special law providing for it other than in Clay's *Digest*, Sec. 10, p. 473, which states that "the mode of trial of slaves should be the same as that for white persons".

[72] Hitchcock, *The Alabama Justice of the Peace*, p. 426; Toulmin's *Digest*, p. 631; Aikin's *Digest*, p. 391.

[73] *Acts of the General Assembly*, 1839, p. 12; Clay's *Digest*, p. 546.

[74] *Penal Code of the State of Alabama*, 1841, Ch. III.

Dr. H. V. Wooten, a prominent physician of Autauga County, made an important notation in his diary on February 10, 1837. On that day he attended a coroner's inquest on the body of a negro boy who the jury thought was murdered by his master, Henry M. Hall. In April Dr. Wooten attended court in Autauga County as a witness in the case. Hall was convicted of manslaughter, imprisoned for two months, and fined $500.[75] In the case of *The State* vs. *Flanigan,* an overseer for Robert Freeman of Jackson County was charged with whipping to death a negro slave named Jacob. The jury found the prisoner guilty of murder in the second degree, and he was sentenced to ten years imprisonment in the penitentiary.[76]

The Alabama law was specific in stating the duties and obligations of a master toward his slaves. He was required to give his slaves sufficient food and clothing for comfort, and should he refuse, such slaveholder would be subject to indictment and conviction and a fine of not more than $500 by the jury judging the case.[77] The Act of 1848 placed on the jury trying such cases the sole responsibility for determining whether or not the food and clothing were "sufficient and necessary".[78] In the event of continued neglect or refusal to comply with the laws of the State, the General Assembly had the power to sell the slaves of a remiss owner. However, the owner was to receive the proceeds of the sale.[79]

In the case of *Fisher and Johnson* vs. *Campbell* (1839), the court held that the owner was bound by the action of his overseer in providing food for his slaves. The defendant, who lived in South Carolina, owned a plantation in Lowndes County, Alabama, which was managed

---

75 *Private Journal of Dr. H. V. Wooten,* Book I, 104-105.

76 5 Ala. 477; *Turnipseed* vs. *The State,* 6 Ala. 664.

77 *Acts of the General Assembly,* 1832, p. 18.

78 *Alabama Beacon,* June 10, 1848.

79 Clay's *Digest,* p. 539.

by an overseer named McMoy. On one occasion the food for the negroes ran low and the overseer bought eight barrels of pork from Fisher and Johnson, with whom Campbell had previously done no business. The price charged compared favorably with the quotations of other grocery concerns in Montgomery, but the defendant refused payment. The Supreme Court held that McMoy was really an agent of the owner, who was liable to pay. An overseer could not bind the master to pay for unnecessary items used on the plantation, but food for the slaves was held necessary.[80]

In 1856, a law was passed stating that not more than six negroes could reside on any plantation in the State of Alabama unless the owner, overseer, or white agent remained on the same place with them or within the distance of a mile. Any violation of this law subjected the owner to a fine of $100.[81] This statute performed the double function of keeping the negroes in order and preventing undesirable persons from loitering around with the hopes of enticing the negroes to leave or actually stealing them.[82]

Numerous cases arose regarding the hiring of slaves. The law held the hirer responsible if the slave was injured or killed due to carelessness or neglect. Such a case arose when a negro who was hired on a river boat was killed because of the negligence of the owners, who had failed to protect sufficiently well the moving machinery. The court held that death was in no way due to the negligence of the slave. Therefore, his owner could re-

---

[80] 9 Porter 210.

[81] *Acts of the General Assembly*, 1855-1856, p. 18; *Molett* vs. *The State*, 33 Ala. 409.

[82] *Mooney* vs. *The State*, 8 Ala. 328; *Nabors* vs. *The State*, 6 Ala. 200; *Prince* vs. *The State*, 3 Stewart and Porter 253. "Any person who shall inveigle, steal, or carry away a slave upon conviction shall be confined to the penitentiary for not less than ten years." *Acts of the General Assembly*, 1843, p. 41.

cover damages.[83] The owner of a slave was not bound to pay for medical treatment when a physician was employed by the hirer to treat him,[84] but the person for whom the slave was working was required to provide the necessary medical attention.[85]

The case of *Hooks* vs. *Smith et al.* (1850), involved a slave who was hired as a house servant but was put to work in the field, where he lost his life. Judge Story ruled that "there is on the part of the hirer an implied obligation not only to use the thing with due care and moderation, but also not to apply it to any other use than for which it is hired," and he further stated that if the thing was used for a different purpose than that for which it was intended by the parties, the hirer was responsible for the damages.[86] A somewhat similar situation occurred in the case of *Seay* vs. *Marks* (1853). This case involved a slave named King who was rehired to a third party to raft lumber down the Alabama River. While pursuing this work he was drowned, and the court held that Seay was liable because he did not exercise ordinary care in re-leasing him to do dangerous work.[87]

False warranty for soundness of mind and body in any transfer of slaves allowed the plaintiff to recover under the laws of Alabama.[88] The case of *Athey* vs. *Olive* (Pike County, 1859) arose after a slave woman named Matilda was bought by the plaintiff for $1,100. She was purchased as a household servant but appeared to have capabilities only as a field hand. The plaintiff sued for recovery, but Judge R. W. Walker laid down the following ruling:

---

[83] *Williams and Hitchcock* vs. *Taylor*, 4 Porter 234; *Ala. and Tenn. Rivers Railroad* vs. *Burke*, 27 Ala. 535.

[84] *John Meeker* vs. *James Childress* (1823), 1 Minor 109.

[85] *Hogan* vs. *Carr and Anderson*, 6 Ala. 471.

[86] 18 Ala. 338.

[87] 23 Ala. 532.

[88] Helen T. Catterall, *Judicial Cases Concerning Slavery*, III, 137; *Caldwell* vs. *Wallace*, 4 Stewart and Porter 282.

> Where a slave is neither insane nor an idiot, nor subject to any mental derangement which interferes with the natural operations of the mind, the mere fact that she lacks ordinary sense does not establish a breach of warranty that she is of sound mind. Such a warranty is not a stipulation against mental weakness or a guaranty that the slave is possessed of ordinary capacity.[89]

In Alabama, as elsewhere, slaves were considered as property, as chattels personal, and as such were subject to the laws relating to property. Consequently they were liable to be seized by creditors of their owners and sold to satisfy the debt. This was true in all Southern states except Louisiana, where slaves were held as real estate and could not be separated from the land.[90] The law providing that a creditor could seize a slave for debt of the master perhaps worked a greater hardship upon the slave than the law giving the master the right to dispose of him at will. For although the master may not have desired to dispose of his slave, considering him a valuable piece of property, and may have promised never to sell him from his family, nevertheless the slave could be seized at any time and be made to suffer for the misfortune of his master.

Since slaves were subject to ordinary property laws, few special statutes in regard to them were enacted. Those, in most instances, were in the general nature of protection for the slave. In 1807 the Mississippi Territorial Legislature enacted a law which continued in force many years after Alabama was admitted to the Union. In essence it provided that there should be no levy against slaves for debt unless the amount exceeded $100, and then only after other properties of the planter had been disposed of. In the meantime, if a writ of execution had been served and the slave was not im-

---

[89] 34 Ala. 711.

[90] William Goodell, *The American Slave Code in Theory and Practice*, p. 64.

mediately restored to the owner, sufficient sustenance had to be provided by the sheriff until the negro was sold or otherwise legally discharged.[91]. In addition, if a tax collector levied on the slave property of any defaulter for taxes, the sale of the negroes should not be for a longer period than one year, the right of redemption being reserved to the owner for that time.[92]

The Act of 1815 was practically a reënactment of the same law, providing that slaves were not to be levied on until other property was sold and that names of slaves sold under execution were to be recorded in the records of the court by which the execution was issued.[93]

Only one other special statute appears in the legislation of Alabama humane law enacted after 1843 and before 1852. That law provided that:

> No execution can be levied on a child or children, under the age of ten years, without including the mother, or upon the mother without including the child or children, as aforesaid, if living and belonging to the defendant in execution; and the mother and child, or children, must be sold together, unless the parties in interest, or one of them, make affidavit and deliver the same to the officer, that he believes his interests will be materially prejudiced by selling the slaves together, when they may be sold separately; but no levy or sale shall be made, by which a child under five years of age shall be separated . . . . In all sales of slaves under any decree, or order of the chancery or probate court, or under any deed of trust, or power of sale in a mortgage, the slaves must be offered, and if practicable sold in families. . . .[94]

In the eyes of the civil law, the free negro in Alabama was in many respects on a plane with the free white man. He enjoyed the rights of property and protection, he could contract and be contracted with, sue and be

---

91 Toulmin's *Digest* (1807), pp. 85-86; Clay's *Digest*, pp. 202-203.

92 Toulmin's *Digest* (1807), p. 50.

93 Clay's *Digest*, p. 203.

94 *Alabama Code of 1852*, Secs. 2056-2057.

sued, inherit and bequeath property; his marriage was legal; and he could mortgage his property. He was *sui juris* and in civil law enjoyed the privileges of a free man. In criminal law his position was not quite the same, for many acts which the white man could commit with impunity were set down as crimes for him.

There was among slaveholders a very general impression that the influence of a free negro upon slaves was demoralizing. The slave might become restless under the taunts of his free brother or, encouraged to hope for freedom himself, might run away. Hence, after 1834, the freed slave was forced to leave the State within twelve months.[95] There seems to have been a current idea in Alabama, as well as in the South in general, that unrestrained manumission would lead to one of two conditions: either the free negro would remain in competition with the white laborer, or he would become "lord of the soil" as he had in Haiti and Jamaica. Most Southerners believed that the free negro was incapable of providing sustenance for himself in competition with the whites, and that a large percentage of negroes without steady work would resort to pillage and theft. The alternative of their becoming owners of the soil was even more to be avoided.

Until Alabama became a state, no laws governing free negroes existed, and not until 1832 was there any statute forbidding the immigration of free persons of color into the State.[96] Section 9 of the Act of 1832 stated:

> And be it further enacted, that from and after the first day of January next it shall not be lawful for any free person of color to settle within the limits of this State; and should any free person of color, after that time, settle in this State, he, she, or they shall on notice of this act, depart within thirty days or shall be liable on conviction before any justice of the peace to receive thirty-nine lashes; and any person may arrest any such free per-

[95] Toulmin's *Digest*, p. 646; *Atwood's Heirs* vs. *Beck*, 21 Ala. 590.

[96] *Acts of the General Assembly*, 1831-1832, p. 188.

> son of color and take him or her before any justice of the peace for trial. . . .[97]

The laws that dealt with the free negro in his relations with the slave were not otherwise than should be expected, for having come to depend on slavery as a means of livelihood, the people of Alabama did not care to have it interfered with and desired to preserve good order and contentment among the negroes. They believed the free negro was injurious to their slave interests, and that the less contact between the two elements the better. Thus, the free negro was punished for dealing with a slave, aiding him to escape, visiting a plantation without the permission of the owner, preaching without a license, writing a pass, retailing liquor or buying or selling to a slave who was unauthorized by his master.[98] In 1848, the Supreme Court even went so far as to lay down a ruling that all cotton samples used by commission merchants had to be taken by free white persons. Any factor who authorized such samples to be taken by a free negro was subject to a fine.[99]

The Constitution of Alabama proclaimed in 1819 contained the following statement in regard to emancipation:

> The General Assembly shall have no power to pass laws for the emancipation of slaves, without the consent of their owners, or without paying their owners, previous to such emancipation, a full equivalent in money for the slaves so emancipated . . . . They shall have power to pass laws to permit the owners of slaves to emancipate them, saving the rights of creditors, and preventing them from becoming a public charge.[100]

---

[97] *Ibid.*

[98] *Acts of the General Assembly,* 1822, p. 61; *ibid.,* 1832, p. 16; Aikin's *Digest,* p. 397; *The State of Alabama* vs. *McDonald,* 4 Porter 449.

[99] *Wragg* vs. *The State,* 14 Ala. 492.

[100] *The Federal and State Constitutions, Colonial Charters and other Organic Laws of the United States,* (compiled by B. P. Poore), Part I, p. 44.

The early statute law regarding manumission in Alabama was an act of the Mississippi Territorial Legislature, passed in 1805, and remained the law until 1834. This act provided that it was not lawful for any person owning slaves to liberate them until they proved to the satisfaction of the General Assembly that such slave had done some meritorious act for the benefit of the owner or the Territory.[101] In addition, the owner of the slave had to give security to the governor that the slave, at least for the time being, would not become a public charge. Therefore, manumission in Alabama was possible only by a general or special act of the legislature.[102]

The law was specific in 1834 in stating that the slave who was freed by his master should move beyond the limits of the State within twelve months and that complete manumission should not take place until after such removal. If the slave should return after his removal and manumission, he was to be apprehended by the sheriff of the county in which he was found and imprisoned. After this fact had been advertised for at least thirty days, he could be sold to the highest bidder for cash as a slave for life, and the proceeds of such sale, after all expenses were paid, were to go to the county treasury for the use of the county.[103] The law did not provide for the expulsion of free negroes who were already in the State.

Another noticeable feature of this statute is that it contained nothing to prevent a person from directing in his will that slaves be taken out of the State by his executors to a free territory and there manumitted.[104]

---

[101] See Mississippi Constitution 1817, *ibid.*, Part II, p. 1064; Toulmin's *Digest*, p. 632; *Trotter* vs. *Blocker*, 6 Porter 269.

[102] Toulmin's *Digest*, p. 644; *Acts of the General Assembly*, 1821, p. 100; Catterall, *Judicial Cases Concerning Slavery*, III, 135; *Roberson's Heirs* vs. *Roberson's Executors*, 21 Ala. 273.

[103] *Acts of the General Assembly*, 1834, p. 29; Clay's *Digest*, p. 454.

[104] *Hooper* vs. *Hooper*, 32 Ala. 669.

However, in 1860 this was changed by the following law: "That from the passage of this act, all clauses, items, words or directions, contained in any last will and testament or ordinary instrument made for the purpose of emancipating any negro slave or slaves, either directly or indirectly is hereby declared void and of no effect."[105] Thus, as the slave regime came to an end, it became virtually impossible to manumit by will and testament. Although rather liberal in its policy of passing special manumission laws in the twenties and thirties, the legislature became less and less inclined in this direction. Only through meritorious service which was nothing short of heroism could a slave obtain his freedom.

---

105 *Acts of the General Assembly*, 1860, p. 28.

CHAPTER VI

TRANSPORTATION AND EXPORT OF COTTON

COTTON was the key to the whole economic situation in ante-bellum Alabama. Planters not only needed adequate means of sending their product to the seaboard, but they also demanded manufactured articles from the North or from Europe, and corn, meat and flour from the Northwest. Therefore, internal improvements were necessary, particularly in the cotton planting areas. On the other hand, the inhabitants of the mountain region and wiregrass district produced for themselves nearly all of their required commodities. They were more largely independent of communications abroad and later created little effective demand for railroads. Nearly all parts of Alabama were supplied with natural waterways giving outlets in some direction, and it was these rivers which became the principal arteries of trade between the interior and the coast.[1]

Before the advent of the steamboat in the early twenties, cotton was floated down to market on flatboats. As early as 1822 the Taits were searching for a suitable barge to transport their cotton from Wilcox County to Mobile. James wrote his father, "It will be proper to engage a barge to carry down our cotton. I'll be glad if you will look out for one. Any time after the first of March I can have it at the landing in three days."[2] Flatboat navigation, which necessitated great familiarity with the rivers, became a sort of profession. In some

---

[1] Ulrich B. Phillips, *The History of Transportation in the Eastern Cotton Belt to 1860*, pp. 8-9.

[2] *Charles Tait Papers*, Feb. 1, 1822.

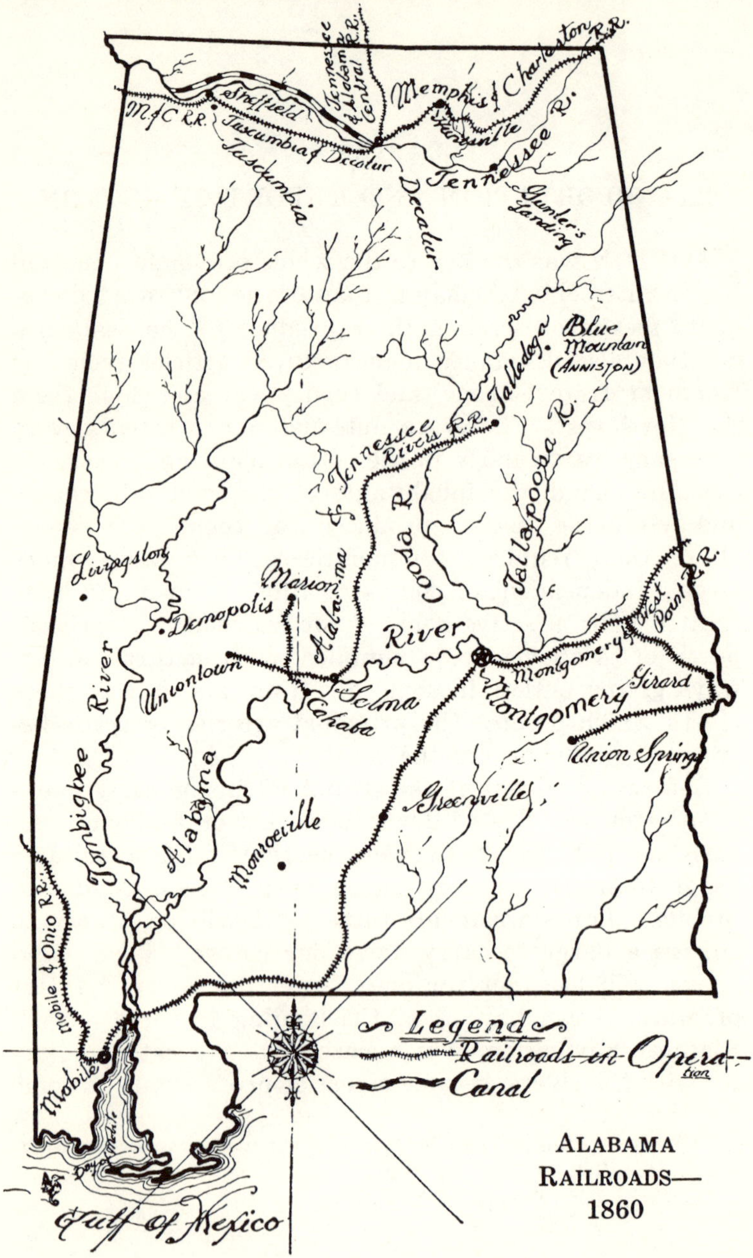

ALABAMA
RAILROADS—
1860

instances, the boatmen agreed to sell the cotton after it arrived at the export city and return to the owner the proceeds of the sale after having deducted certain charges such as freight, insurance, storage and commissions.[3]

The average length of time required by a keel boat to reach New Orleans from the Tennessee Valley was about four months, and frequently factors complained that the length of time involved prevented them from giving efficient service due to a rapidly changing market.[4] The flatboats often traveled in fleets. In the spring of 1822 thirty-two boats loaded with 8,268 bales set out from Ditto's Landing on the Tennessee River.[5] Nelson and Davis of Huntsville and White and English of Florence seem to have enjoyed the largest barging business to New Orleans. The former owned warehouses at Cottonport Landing and on the south side of the river at Nelson's Ferry and advertised not only first-rate boats and experienced personnel but also free storage for customers who shipped on their boats.[6]

Muscle Shoals presented a serious barrier for shipping from the upper Tennessee. It was always necessary for cotton to be held until a spring freshet made it possible for experienced pilots to guide the boats safely over the shoals where they were delivered to their owners and the pilots returned to carry down more boats.[7] Frequently, cotton was damaged by mud and water en route to market and consequently brought a lower price. A New York mercantile house in 1829 complained that

---

[3] *Alabama Republican*, Jan. 19, 1821; T. J. Campbell, *The Upper Tennessee*, p. 7.

[4] Manusel White to John Coffee, March 9, 1826, *John Coffee Papers*; *History of Navigation on the Tennessee River System*, House Doc. 254, 75th Cong., 1st Sess., Ch. VI.

[5] *Alabama Republican*, May 24, 1822.

[6] *Ibid.*, Jan. 26, 1821; *John Coffee Papers*, Jan. 7, 1826.

[7] *Huntsville Democrat*, Jan. 5, 1827; John R. Kennamer, *History of Jackson County*, p. 28.

North Alabama cotton not only arrived in damaged condition but was also poor and trashy, and the merchants inferred that it had been raked from the ground instead of having been picked. Their agents in New Orleans were ordered to buy Mississippi and Louisiana cotton whenever possible.[8]

In order to connect Huntsville with the Tennessee River, Big Spring branch was made navigable by a series of locks and dams to its confluence with Indian Creek. The company, known as the Indian Creek Navigation Company, in which stock was sold, was chartered by an act of the legislature in 1820.[9] Progress was slow, but by 1827 the canal was within two and one-half miles of Huntsville and during that winter several thousand bales were floated down to the river in light-draft boats. Formerly, planters had been forced to pay a dollar a bale to have their cotton hauled to Triana, but by the canal the same could be accomplished for twenty-five cents. This canal was the first in the State, and the *Southern Advocate* was impressed by the fact that "our citizens who have never been on a canal are gratified at the novel spectacle of seeing boats heavily laden with cotton, pass through a lock, descending several feet in as many minutes, without the least risk or danger".[10]

Even on the smaller rivers, flatboats transported cotton to market and sometimes were poled back against the current laden with plantation supplies. An instrument familiarly known among the boatmen as a "hook and jam" was indispensable in moving the clumsy barges upstream. It consisted of a long smooth pole, pointed with a long spike, and a hook curving from its beak but a few inches from the point. The boat was moved by pressing the point against trees or by fastening the hook

---

[8] *Southern Advocate,* Sept. 11, 1829.

[9] Edward Chambers Betts, *Early History of Huntsville, Alabama,* p. 66.

[10] *Southern Advocate,* Jan. 27, 1827.

to overhanging boughs.[11] At St. Stephens, on the Tombigbee, one frequently noticed barges from forty to fifty feet in length loaded with several hundred bales of cotton floating down the river on their way to Mobile.[12]

Planters who were not located adjacent to the rivers frequently sacrificed a large part of their profits in the cost of transporting the cotton crop to market. Early roads were little more than trails and during the rains of winter and spring were almost impassable for cotton wagons, which were usually drawn by four mules. Buckingham observed that planters were not interested in improved roads since they were always on the lookout for an opportunity to sell their holdings at a profit and move on to new lands.[13] For purposes of road maintenance each county was divided into precincts, and a certain portion of a road was delegated to an individual who was supposed to see that it was kept in a state of repair sufficient for wagon traffic.[14]

Cotton was hauled from the plantation to the nearest river bluff, where the bales were slid down an incline to the waiting steamboat or barge.[15] Every planter on the river had a private landing, and those in the interior shipped either from the landings of their neighbors or from a public landing. Planters in the community of Tilden in south Dallas County hauled their cotton to Portland on the Alabama River, where it was stored in the warehouse of C. Hall and Company while awaiting the arrival of boats from Mobile.[16] In Wilcox County, Bridgeport was the central shipping point not only for the Camden district but even for cotton wagoned from Butler and Conecuh Counties, a distance of forty or

---

[11] B. F. Riley, *History of Conecuh County*, p. 77.

[12] T. H. Ball, *History of Clarke County*, p. 175.

[13] J. S. Buckingham, *The Slave States of America*, I, 258.

[14] *Lauderdale County Road Book*, 1844-1860.

[15] *Dubois Diary*, Nov. 4, 1840.

[16] *Warehouse Ledger of C. Hall and Company*, 1854-1860.

fifty miles from the river. In Greene County, the Gould plantation had access to either Hayes' Ferry or Miller's Landing on the Tombigbee. The latter place was five miles distant and more frequently used except during extremely low stages of the river.[17]

The spring of 1855 was unusually dry, and the rivers became so low that boats were unable to transport cotton to Mobile. Both planters and factors were equally alarmed, and some even considered building flatboats to float their cotton down as they had done in the early days. On March 23, Gould recorded in his diary that two small stern-wheel boats arrived at Miller's Landing but demanded five dollars a bale to take cotton to Mobile.[18]

Though the St. Stephens Steam Boat Company was chartered by the Territorial legislature in 1818,[19] it is almost impossible to fix accurately the date of the first appearance of a steamboat on the rivers of Alabama. Hamilton, in his *Colonial Mobile,* asserts that the first trip as far up as Demopolis was made in 1819, either by the *Tensa* or the *Mobile.*[20] Until October, 1821, no steamboat had come up as high as Montgomery on the Alabama River because of the fear of sand bars, islands and other obstructions in the narrow channels. The feasibility of steam navigation was fully established, however, when the steamboat *Harriet,* commanded by Captain Morril, arrived at Montgomery after a ten-day trip from Mobile.[21] By 1822 there were three steamboats on the Alabama River and one on the Tombigbee carrying an average load of 400 bales of cotton each trip.[22] The

---

[17] *William P. Gould Papers,* 1855; William E. W. Yerby, *History of Greensboro,* p. 10 ff.

[18] *William P. Gould Papers,* 1855.

[19] Toulmin's *Digest,* p. 74.

[20] Peter J. Hamilton, *Colonial Mobile,* pp. 447-448, 471-472.

[21] Mell A. Frazer, "Early Steam Boats in Alabama", *Alabama Polytechnic Institute Historical Studies,* Third Series (Reprint), p. 4.

[22] *Mobile Commercial Register,* June 10, 1822.

first boat to reach Florence on the Tennessee River came in 1821.[23] It was during the next year that a boat called the *Rocket* began a regular run between Florence and Trinity, where cargo was deposited to be forwarded up the Ohio or down the Mississippi on larger vessels.[24]

Shippers above the shoals continued to send their cotton down in keelboats, which were often towed to New Orleans or transferred to other boats. Freight rates which had formerly been one dollar a hundredweight in 1822 continued to fall until 1828, when they became half that amount. Thus New Orleans came to supplant Nashville more and more as a market for plantation supplies.[25]

The size of boats was usually reckoned by the number of bales of cotton carried. Until 1830, from 200 to 400 bales were considered an average load, which was hardly more than the flatboats could carry.[26] In 1859, the *Alice Vivian* arrived in Mobile with 2,087 bales, which was considered an unusually large cargo for a boat plying the rivers of Alabama.[27] The schedule of river boats depended to a large extent upon the condition of the rivers and the amount of freight to be taken on board at the various landings at which the steamer stopped. Lyell observed that a variation of a day in arriving or departing was not uncommon, and he regretted having to wait many hours on the bank of the river for the arrival of the boat.[28] In the early days the trip from Mobile to Montgomery required eight or ten days, but by 1840 this time had been cut to forty hours.[29]

As early as 1821 Governor Pickens called attention to the necessity of a canal by which the Tennessee and Alabama Rivers could be connected. This proposed pro-

---

23 *Alabama Republican*, March 16, 1821.

24 *Ibid.*, Jan. 18, Feb. 1, March 15, 1822.

25 *Ibid.*, Jan. 18, 1822.

26 Frazer, *Early Steam Boats in Alabama*, p. 19.

27 *Clarke County Democrat*, Jan. 27, 1859.

28 Charles Lyell, *A Second Visit to the United States*, II, 44.

29 *Selma Free Press*, April 25, 1840.

ject had long been deemed necessary as a means of more closely uniting the northern and southern sections of the State.[30] The mountain barrier which separated the two sections was penetrated only here and there by trails which were inadequate as highways of commerce. North Alabama was more closely connected with Louisiana than with South Alabama. Products from the northern part of the State were shipped down the Tennessee, Ohio and Mississippi Rivers to New Orleans, a distance of 1,500 miles, and from this place the greater proportion of the plantation supplies was purchased. On account of the shoals in the Tennessee River even this means of transportation was retarded during the dry season, and markets had to be sought in Savannah, Augusta and Charleston, to which places about a third of the distance had to be traversed by wagons.[31]

Governor Gayle, in his annual message in 1834, stated that such a canal uniting the Tennessee and Alabama water systems would carry to Mobile annually 150,000 bales of cotton which were then going to other states by dangerous and more expensive routes.[32] Not only was Mobile being deprived of that share of the State's traffic, but heavy losses were being sustained by planters of North Alabama because of the lack of adequate transportation facilities. For example, in 1833 cotton was worth fifteen cents a pound in New Orleans and Mobile, but before the high water came, admitting passage over Muscle Shoals, the price of cotton had fallen five cents. During this year alone it was estimated that the loss to planters in the Tennessee Valley amounted to over a million dollars.[33]

---

[30] William E. Martin, "Internal Improvement of Alabama", *Johns Hopkins University Studies in Historical and Political Science*, Series 20, No. 4, p. 35.

[31] *Ibid.*

[32] *Ibid.*, pp. 35-36.

[33] *Ibid.*, p. 36.

At the same time that the canal to connect the Hiwassee and the Coosa was being considered, the introduction of steam navigation made it even more evident that a canal around Muscle Shoals was greatly to be desired. A government survey was made and a report submitted, but Congress did not believe it had the right to appropriate money for the construction of such a project.[34] However, there seemed to be no objection to donating public lands to the states for purposes of internal improvements, and in 1828 Congress granted the proceeds from 400,000 acres in the neighborhood of Huntsville to build a steamboat canal around the shoals.[35] The Muscle Shoals canal was considered of much more national importance than the Coosa-Hiwassee project, and the latter, after several futile attempts to put it into operation, was finally abandoned.[36]

Before the canal around Muscle Shoals could be completed, the Tuscumbia, Courtland and Decatur Railroad connecting Tuscumbia and Decatur was constructed.[37] Cotton could then be safely carried around the dangerous shoal waters at a cost of two cents per bale per mile.[38] It frequently happened that flatboats making the same trip by river were obliged to lighten their cargoes at great expense and delay. In order to encourage trade with the railroad only 12½ cents a bale warehouse charges were made on cotton handled at Decatur, provided boats were waiting at Tuscumbia to receive it when the train arrived. It was a well-known fact that cotton above the shoals was from sixty to ninety days longer getting to New Orleans by water than that below.

---

34 *American State Papers, Military*, IV, 13; *Southern Advocate*, March 23, 1827.

35 *Ibid.*, March 20, 1829.

36 Thomas P. Abernethy, *The Formative Period in Alabama*, p. 80.

37 *Huntsville Democrat*, Dec. 31, 1834.

38 *Ibid.*, Feb. 25, 1835.

Freight rates from above Florence were fifty cents a bale more, besides additional charges for river insurance.[39] When the canal was finally put into operation it was not uncommon for spring floods to put it out of commission, as was the case in 1841, when all traffic had to be suspended until repairs could be made.[40]

During the forties, plank roads were experimented with in Alabama, but planters apparently were not interested in paying tolls. The fact that cotton hauling came at a time when other work on the plantation was not urgent seemed to justify a long haul over unimproved roads.[41] One plank road, however, played an important part in the transportation of cotton to market. This road ran southward from Sylacauga to Wetumpka, a distance of seventy miles, and served as a connecting link with the head of navigation on the Coosa River.[42] Owing to obstructions in the Coosa, planters of this section were compelled to wagon their produce south to the nearest shipping point on the river accessible to Mobile, or haul it north to the mouth of Canoe Creek, where it was transported upstream 160 miles to Rome, the western terminus of the Georgia railroads. By this outlet a great portion of trade of the middle, eastern and northeastern counties of Alabama was diverted from Mobile, notwithstanding the fact that the frequent shifting and handling of freight and the distance to Charleston and Savannah caused the freight rate to total $4.30 a bale. During the year 1850 it was estimated that no less than 10,000 bales were drawn away from Mobile by this route.[43]

Much cotton, however, from Talladega County found

---

39 *Ibid.*

40 *Ibid.*, Jan. 30, 1841.

41 Phillips, *Transportation in the Eastern Cotton Belt*, p. 12.

42 *Sumter County Whig*, Aug. 17, 1852.

43 *DeBow's Review*, Aug., 1850, p. 220.

its way to Wetumpka over the plank road.[44] It was generally asserted that a team of mules could pull three times the load on a road constructed of plank that they could on one of dirt, and at a considerably greater speed.[45] The fastest trip over the plank road between Montgomery and Sylacauga appears to have been made by William L. Yancey on July 4, 1852, when he made a round trip between the two towns in fifteen hours, which also included dinner and a short speech.[46]

In western Alabama, Cahaba, Demopolis and Erie were three early important cotton-shipping points. The first-mentioned town, for a short time the seat of government in Alabama, was admirably located to serve Dallas and parts of Perry and Lowndes Counties.[47] Cotton grown in eastern Greene County in the vicinity of Greensboro was wagoned to Erie, where it was shipped to Mobile by river at the rate of a dollar a bale.[48] However, the prairie roads offered an almost impassable barrier, especially during the heavy winter rains.[49] Planters living west of Uniontown in Perry County hauled their cotton to Demopolis for shipment. In seasons of low water on the Warrior more cotton came to the Tombigbee than to the Alabama, and in 1860, 25,000 bales were forwarded through Demopolis to Mobile.[50]

Not only did Alabama cotton pass out of the State in the northeast, but considerable amounts were sold through Appalachicola and Pensacola. Randolph, Chambers, Russell and Barbour Counties had ready access to the Chattahoochee River, which made possible for

---

44 *Matthew P. Blue Papers*, Sketches of Talladega County; Alderman and Gordon, *J. L. M. Curry — A Biography*, p. 40.

45 *Alabama Beacon*, Oct. 27, 1849.

46 *Sumter County Whig*, Aug. 17, 1852.

47 John W. DuBose, *The Life and Times of William L. Yancey*, p. 77.

48 *Greene County Gazette*, Dec. 20, 1830.

49 R. T. Neighbors, *Plantation Life in Hale County*, p. 45.

50 J. W. DuBose, *Chronicles of the Canebrake*, p. 2.

them markets in Florida for both selling cotton and purchasing plantation supplies. The Conecuh River was an important waterway for South-Central Alabama. As early as 1823 cotton was being floated down to Pensacola Bay, where tonnage was paid at the customs house and licenses were obtained for steering the boats up to the wharves of the city.[51]

In the Tennessee Valley considerable cotton was passing up the Ohio River. H. L. Perry of Cincinnati was a frequent customer of General John Coffee and was continually complaining about the poor grade and damaged condition of the cotton when he received it.[52] In 1828 Warfield Brand of Lexington, Kentucky, offered to buy Coffee's entire crop of seventy bales at a price of 10½ cents a pound delivered at Louisville any time between the months of November and April. Coffee, however, after much deliberation turned down the offer, hoping to get a better price in New Orleans.[53]

The appearance of railroads in Alabama materially affected the traffic on the rivers, although before 1860 the railroad was never able to supplant entirely its older rival. In 1853, river traffic between Montgomery and Mobile amounted to 250,000 bales of cotton, 460,000 barrels of merchandise and 30,000 passengers, the receipts from which amounted to $631,000.[54] From Whitesburg, the shipping point for Huntsville, to Charleston the freight rate was $6.38 a bale by way of the Tennessee River to Chattanooga and thence over the Georgia railroads.[55] C. C. Clay, Jr., found that certain commission

---

[51] Riley, *History of Conecuh County*, pp. 48, 78.

[52] *John Coffee Papers*, March 24, 1821.

[53] *Ibid.*, Nov. 24, 1828.

[54] Albert Stein, *Report on the Improvement of the Alabama River*, 1855.

[55] *David E. Kelley Papers; Reports of the President, Directors and Chief Engineer of the Alabama and Tennessee Rivers Railroad Company*, 1851, p. 51.

houses in Savannah charged only fifty cents a bale for handling cotton, which was less than half the usual commission of 2½% in Charleston.[56] The factorage house of Thomas, Wilson and Company was quite reliable and the Savannah market equally as good as that at Charleston.[57] The development of Chattanooga as a railroad center by 1850 tended to break the monopoly held by New Orleans as an exclusive market for Tennessee Valley cotton.[58]

As early as 1846, newspapers of North Alabama were urging planters in the northeastern counties to ship their cotton up the Coosa River instead of wagoning it over the Central Plank Road to Wetumpka.[59] Before the completion of the Western and Atlantic Railroad into Cherokee County, Mobile had been the principal market for cotton from this part of the State. North Alabama planters failed to secure the prestige in factorage accommodations held by those of the Black Belt and were usually forced to sell at any price the buyers might choose to give.[60] However, with an outlet up the Coosa River, Charleston, Augusta and Savannah, as well as Mobile, offered facilities not only for marketing cotton but also for securing advances of capital which were always necessary for planting.[61]

In 1850, 350,952 bales of cotton were received at Mobile, of which 187,130 came down the Alabama River, 101,208 down the Tombigbee, 56,276 via the Warrior, and 6,314 by wagons.[62] Considerable cotton appears to have been wagoned into Mobile from eastern counties

---

56 Thomas, Wilson and Company to C. C. Clay, Jr., Feb. 10, 1851, *Clay MSS.*

57 Same to same, *ibid.*

58 *J. R. Mason Papers*, 1850; *Southern Advocate*, March 13, 1846.

59 *Ibid.*, Nov. 12, 1846.

60 *Ibid.*, Jan. 30, 1846.

61 *Ibid.*, Jan. 30, Nov. 12, 1846.

62 *DeBow's Review*, Dec., 1850, p. 657.

of Mississippi. Ox teams consisting of six or eight yoke of steers frequently hauled twenty-five or thirty bales at a load and required several days to make the trip.[63] In 1851, over 18,000 bales were shipped direct from Montgomery to New Orleans and passed through Mobile without being included in the receipts for the year.[64]

In attempting to appraise the value of railroads in the transportation of cotton in Alabama it is necessary to consider their development during the ten years prior to the War. In 1851 the railroads of the State were confined to the Montgomery and West Point in the direction of Atlanta, a fragment of the Charleston and Nashville road, the Mobile and Ohio, the Alabama and Tennessee Rivers Railroad and the Blakely and Girard.[65] In its effort to reach the Ohio River, the Mobile and Ohio had completed 321 miles of track, 164 of which were in Alabama. It was 1859 before this road was actually completed to Cairo, Illinois, the proposed terminus.[66] The citizens of Mobile voted a tax on their real estate to the sum of $300,000 in order to hasten the construction of the road and thus receive the benefits of an increased trade with the interior. Planters along the route were in nearly every case a considerable distance from river navigation, which usually meant that their cotton was at least two months late getting to market. The railroad agreed to carry cotton from points in Alabama and Mississippi to Mobile at from $2.50 to $3.00 a bale.[67]

The second important railroad in the State was the Alabama and Tennessee Rivers Railroad, commencing at Selma and extending to Gunter's landing on the Ten-

---

63 J. F. H. Claiborne, "Trip through the Piney Woods", *Publications of the Mississippi Historical Society*, IX, 531.

64 *DeBow's Review*, Dec., 1851, p. 647.

65 *Ibid.*, Aug., 1851, p. 161.

66 Harold W. Faulkner, *American Economic History*, p. 345.

67 *DeBow's Review*, Aug., 1851, p. 161.

nessee River, from which point proposed branches were to run to Chattanooga and Rome. Thus the dream of connecting the two great waterways of Alabama was accomplished. The proposed route of the Blakely and Girard was from Blakely, on the eastern side of Mobile Bay, to Girard, opposite Columbus on the Chattahoochee River. This would allow cotton from South-Central Alabama to be brought directly to Mobile or be shipped down the Chattahoochee to Appalachicola.[68]

Since the upper waters of the Alabama rivers could not be counted on as a reliable year-round medium of transportation, a railroad from Uniontown in Perry County to Port Jackson on the Tombigbee River was considered in 1858. Such a road would connect the western part of the Black Belt with Mobile, and make possible a certain and reliable means of transportation no matter what the stage of the river might be. The proposed road was to run through one of the richest planting sections of the State, and its terminus would be at the head of tidewater navigation, which would insure year-round transportation.[69]

The construction of the Northeast and Southwest, Selma and Gulf, and Uniontown and Jackson Railroads brought apprehension to Mobile factors and merchants that the cotton export trade might be diverted by way of Charleston and Savannah. If a railroad should have been built south from Selma to intersect the Blakely and Girard road at Pensacola, it might have become a means of sending cotton from twenty Alabama counties out by way of Florida to foreign markets. It was generally believed in Mobile that the citizens of Selma were backing the scheme as a dry-weather conveyance in the hope of sending all cotton which should come in by way of the Alabama and Tennessee Rivers Railroad and the

[68] *Ibid.*, p. 162.

[69] *Clarke County Democrat*, Nov. 11, 1858.

Uniontown and Selma to Pensacola, the nearest shipping point.[70]

For three successive years prior to 1860 the Mobile and Ohio Railroad hauled over 100,000 bales of cotton into Mobile. The greatest business during a single month of the period was in November, 1859, when 50,977 bales were transported from Tennessee, Mississippi and Alabama.[71] During the year 1860, at an average freight rate of $2.29 per bale, 179,764 bales brought the railroad $412,295.20, or 42½% of its total earnings.[72]

The maritime position of Mobile at the mouth of the great Alabama river system was an ideal location for a great cotton export city. Cotton was the circulating blood which gave life to the community. All citizens were interested in the staple, and a failure of the crop brought disaster to the princely merchant and the lowly compress hand alike.[73] So pronounced was the interest in the buying, selling and exporting of cotton that a traveler in the fifties described the city as follows:

> Mobile is a pleasant cotton city of some thirty thousand inhabitants — where people live in cotton houses and ride in cotton carriages. They buy cotton, sell cotton, think cotton, eat cotton, drink cotton, and dream cotton. They marry cotton wives, and unto them are born cotton children. In enumerating the charms of a fair widow, they begin by saying she makes so many bales of cotton. It is the great staple, the sum and substance of Alabama. It has made Mobile and all its citizens.[74]

---

70 *Ibid.*, Dec. 16, 1858.

71 *DeBow's Review*, Nov., 1860, p. 655.

72 *Proceeding of the Twelfth Annual Meeting of Stockholders of the Mobile and Ohio Railroad*, 1860.

73 J. H. Ingraham, ed., *The Sunny South or the Southerner at Home*, p. 508; James Stuart, *Three Years in North America*, II, 123.

74 Hiram Fuller, *Belle Brittan on a Tour*, p. 112.

Up to the time of the War, Mobile was typically a cotton export city and in 1860 shipped more cotton than any other city in the country with the exception of New Orleans.[75] A good part of her population was floating, since many planters and cotton buyers came and went with the season. During the busy fall and winter steam compresses worked overtime preparing the bales for export, and the constant rattle of the drays carrying cotton between the press and the docks became a familiar sound. Brokers rushed from classing-room to classing-room and from bar-room to bar-room, transacting business with equal rapidity at both places. They were just as likely to receive an order at one place as at the other. The nights brought balls, theaters and supper parties and the days buying, weighing and sampling of both cotton and liquors.[76] Cotton yielded employment to everybody, and during the six months busy season brought in enough income to allow many to spend the rest of the year in idleness. What the spendthrift could not make, he could usually arrange to borrow with the promise of principal and interest when the next busy season came again.[77]

During the twenties, Blakely, on the eastern side of Mobile Bay, probably enjoyed a more favorable position as a port than her rival on the western side because of deeper water on the eastern side which allowed larger vessels to come up the bay.[78] In 1828, John Grant dredged Pinto Pass, which allowed vessels of about the same size to come to Mobile as went to Blakely, and it seems quite probable that this fact was partially responsible for the gradual death of Blakely and the proportionate growth of Mobile. However, during the whole ante-bellum period,

---

[75] Paul Ravesies, *Scenes and Settlers of Alabama*, p. 37.

[76] *Ibid.*, pp. 36-37.

[77] *Ibid.*, p. 38.

[78] Thomas M. Owen, *Publications of the Alabama Historical Society*, III, 115-116.

ocean-going vessels of any considerable size could not reach the city because of the shallow condition of the bay. The great fleet of cotton ships usually rode at anchor twenty-seven miles below the city in the lower bay, and all cargoes had to be transported to and from there by lighters at an estimated annual cost of not less than $100,000 a year. In passing up and down the bay, cotton and other goods were "liable to damage from exposure to weather and it is fair to suppose that it was a prominent reason for the Liverpool cotton merchants assuming, as they did, that cotton going by way of New Orleans arrived in better order, and so should bring a better price than when it went by way of Mobile".[79]

During 1826 and 1827 there were eight separate appropriations totaling $228,838 made by Congress to improve the harbor of Mobile. The greater part of this amount was spent in that part of Mobile Bay known as Choctaw Pass and Dog River Bar, where a channel 200 feet wide and ten feet deep was obtained. Still, however, the channel to the lower bay was deep enough for only light-draft vessels.[80] Despite this handicap Mobile enjoyed continued prosperity and growth. Between 1814 and 1850 the assessed value of real estate grew from $198,000 to $8,577,025, and the city increased in population from approximately 500 to 20,000 persons.[81]

The largest factory demand for Southern cotton was in the region around Manchester, England, and it was here that the price in the ultimate market was fixed. For the whole mill area around Manchester, Liverpool was the importing city. Here were established the great cotton houses that bought and held the stocks to be

---

[79] Martin, *Johns Hopkins University Studies in Historical and Political Science*, Series 20, No. 4, p. 54.

[80] *Ibid.*, p. 55; *Memorial and Proceedings of the River and Harbor Improvement Convention*, 1885, p. 35.

[81] *DeBow's Review*, Sept., 1851, p. 388; Oct., 1851, p. 527.

sold to the manufacturers as raw cotton was needed. To these great importing houses even the New England mills themselves looked to know the price of the staple. As cotton rose and fell in Liverpool, it rose and fell in New Orleans as soon as the slow mails of the day could carry the information. The purchasing cycle of cotton was, therefore, as follows: the overseer and slaves produced it for the planter, the planter sold it to his commission merchant, who had orders placed with him by a broker in the employment of a buyer working in the interest of a foreign importing house. It quite frequently happened that foreign spinners employed American buyers who worked the year round in the interest of their clients.[82] One of the most successful of this type was Harleston Broun of Broun, Beguin and Company, who represented French mill interests.[83]

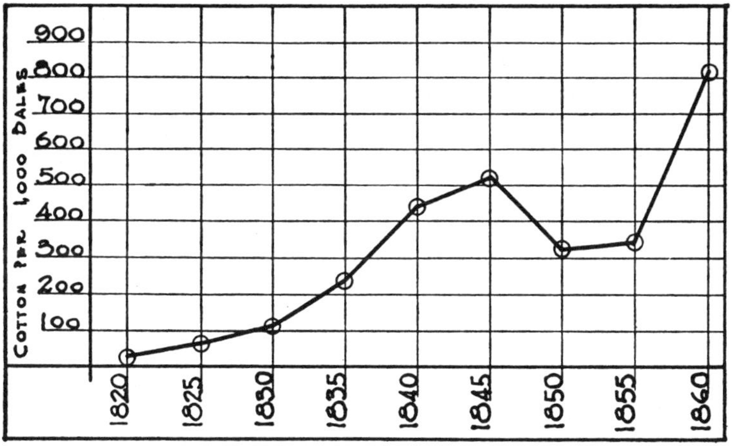

EXPORT OF COTTON FROM MOBILE

[82] John Spencer Bassett, *The Southern Plantation Overseer*, p. 225; *American Cotton Planter*, New Series, II, 159-161.

[83] Information contributed by a descendant of Harleston Broun, Mobile, Ala, 1936.

As early as 1817, when Mobile was yet a struggling village of some 800 inhabitants, 7,000 bales of cotton were exported to eastern manufacturing towns and to Europe.[84] By 1821, 44,000 bales were shipped from Mobile, which was an increase of over 600%.[85] The total crop of the State was estimated at 71,000 bales, which meant that almost half passed out to market by way of other cities. During the same year New Orleans shipped over 100,000 bales, the greater part going to England and France.[86].

During the spring of 1822 a scarcity of vessels greatly handicapped the shipping of cotton. As late as May only one-fourth of the crop had been shipped, and cotton destined for northern ports was sent to New Orleans to be transhipped on the mail packets.[87] This state of affairs led not only to an increased rate of exchange but also to rapidly rising freight rates, because merchants were offering to pay 2½ cents a pound for transportation to Liverpool and two cents to New York.[88] Between the months of January and October, 257 vessels cleared from the port of Mobile, but of this number scarcely twenty sailed directly for foreign ports.[89]

In 1823, records for the shipping season show, 12,934 bales were shipped to foreign markets directly from Mobile, and 31,500 were transported in coastwise trade. Taking into account the 4,157 bales shipped from Blakely, exports from Mobile Bay amounted to nearly 50,000 bales, which was an increase of 5,000 over the year before.[90]

It was discovered in 1823 and 1824 that immense quantities of Alabama cotton were being shipped by

---

84 *Mobile Commercial Register*, May 2, 1822.

85 *Ibid.*; *Niles Register*, XXIII, 131.

86 *John Coffee Papers.*

87 *Mobile Commercial Register*, May 2, 1822.

88 *Ibid.*, April 4, May 9, 1822.

89 *Ibid.*, Jan. 1 - Nov. 1, 1822.

90 *Ibid.*, Oct. 2, 1823.

coastwise trade to New Orleans and from there exported to Europe as Louisiana cotton, for which foreign manufacturers were willing to give a better price. Not only were the additional handling charges added to the cost of the staple, but it was also a well-established fact that exchanges could be more favorably negotiated in Mobile than in New Orleans. By this method of exporting, the poorer grades were being shipped abroad as Alabama cotton. "To the great injury of the reputation and a still greater harm to the pecuniary interests of our State", said the *Mobile Commercial Register,* "our Bay is becoming the mere channel of convenience, to the exporting and importing merchants of New Orleans, and at this rate our merchants will soon become the mere receiving and forwarding agents, and our planters their tributary colonists."[91] Many felt that the only adequate solution of this state of affairs was for the legislature to enact a law which would require the proprietors of gins to put their names and those of the county and State in plain, strong and durable letters on the end and side of each bale, and for the planter to put his name and residence on the opposite end and side. The owners of gins were to register each bale in a book kept for the purpose, with a number and the name of the owner opposite, and all bales would have to be shipped from the State with all the original markings on them, leaving it to the shippers to add such new ones as they might require.[92]

Unsettled political conditions in Europe between 1830 and 1835 brought about a depression in the cotton market and the closing of many mills in England and France.[93] The price of choice cotton dropped to as low

---

[91] *Ibid.,* March 2, 1824.

[92] *Ibid.* Opposition from the planting interests of the State prevented the law from being enacted.

[93] Manusel White to John Coffee, Jan. 29, 1831, *John Coffee Papers.*

as ten cents, and buyers bent every effort to lay in large stores in anticipation of the fact that the market would improve.[94] Foreign exports in the year 1831 indicated that 63,490 bales were bought by Great Britain, 7,423 by France, and approximately 1,000 bales went to other ports in Europe. American spinners, anxious to take advantage of a depressed market, purchased through Mobile over 40,000 bales which went to New York, Boston and other eastern ports as coastwise trade.[95]

By 1835 the world market gave certain indications of improvement, only to be sharply checked by the Panic of 1837.[96] Shipping records of the Port of Mobile indicate rather clearly the large amount of cotton which poured out during the year 1836. The largest single shipment appears to have been carried by the barque *Lady East,* which cleared for Liverpool with a cargo of 600 bales valued at $40,000.[97]

The depression in the cotton market from 1840 to 1845 caused great alarm among the planters, factors and exporters. The price of middling cotton in Mobile reached an ante-bellum low mark of 5¼ cents in 1843.[98] Following the trend of the depressed prices, exports fell off in rapid succession. Over 325,000 bales left the city in 1840, whereas during the next two years the export trade fell off nearly 100,000 bales.[99] A Mobile paper stated, "We have seldom seen our market in so dull and discouraging a condition as at present. The sales are more like those of the summer season than the month of January—averaging some 600 bales per day."[100] In 1841, George Starr of New York wrote Samuel Pickens

---

94 Austill to Tait, Dec. 14, 1832, *Charles Tait Papers.*

95 *Greene County Gazette,* Nov. 24, 1831.

96 Woodbury's Report on Cotton, *House Ex. Docs.,* 24th Cong., 1st Sess., IV, No. 146; *Mobile Mercantile Advertiser,* Nov. 20, 1835.

97 *Reports and Manifests,* April 29, 1836.

98 *Mobile Register and Journal,* March 6, 1843.

99 *Ibid.,* Aug. 10, 1846.

100 *Ibid.,* Jan. 4, 1842.

and urged that cotton be shipped to eastern cities where it would bring a higher price than in the European market. Pickens urged his factor to look into the matter, but since Mobile buyers preferred to do business directly with Europe, the matter was dropped.[101] The ease with which bills of exchange could be secured by exporters seemed to influence their preference for direct foreign trade.[102]

In November, 1844, the stock of cotton on hand in Mobile amounted to 918,000 bales, which was evidence that a great part of the previous year's crop still remained unsold. Many attributed the depression in the Liverpool market to the failure of the potato and grain crops in both Great Britain and on the Continent, but the speculative mania in railroad shares was probably not without its effects. Fortunately for Alabama, however, a severe spring drought throughout most of the South caused a short crop and thus a slight increase in price.[103]

After 1845, market conditions began steadily to improve, even though Liverpool and Manchester spinners kept a firm grip on prices throughout the world. The Southern planter, however, was glad to sell to the "Cotton Lord" at any price which was set.[104] In 1848, Mobile was once more coming into her own, for in that year exports abroad reached the ten-million dollar mark. Of this amount, over two-thirds came directly from Great Britain, while the remainder was divided among France, Belgium, Holland, Prussia, Spain and Sardinia.[105]

In 1850, British spinners consumed 1,200,000 bales of American cotton,[106] 23% of which was exported through

---

101 *Samuel Pickens Papers*, March, 1841.

102 *Minute Book of the Montgomery Branch Bank*, 1838-1843.

103 *Samuel Pickens Papers*, Nov., 1844.

104 *The Southern Cultivator*, VIII, 157; *Alabama Planter*, Jan. 2, 1849.

105 *DeBow's Review*, Dec., 1848, p. 428.

106 *Census of 1850*, Agriculture, p. 191.

the port of Mobile.[107] During the same year Mobile shipped 325,000 bales, which, next to New Orleans, was the greatest aggregate sent from any city in the United States.[108] Between the years 1845 and 1849 the gulf ports of the South shipped over a million and a half bales, while those on the Atlantic coast handled slightly over a half million.[109]

Throughout the late fifties Alabama exports continued to climb upward. Though the average price for middling cotton still fluctuated between nine and eleven cents, planters strained every effort to raise more cotton and in 1860 over 800,000 bales came by river, railroad and wagon to the wharves in Mobile.[110] Slightly over two-thirds of this amount went to foreign markets, bringing a total of $38,333,576. England was the largest customer, consuming 445,663 bales and paying a price of $25,000,000. France was second with purchases totaling $8,559,747, while exports to the rest of Europe amounted to $3,830,760. New England mill owners, though forced to compete with their English rivals with little or no tariff protection, were gradually becoming able to make their importance felt and each year were buying more and more Southern cotton. In 1860 Mobile's coastwise trade with the Northeastern states reached a new high level of approximately 200,000 bales.[111]

---

107 *DeBow's Review*, Dec., 1850, p. 657.

108 *Ibid.*, Aug., 1850, p. 209.

109 *Ibid.*, Nov., 1850, p. 656.

110 *Ibid.*, Nov., 1860, p. 666.

111 *Ibid.*, p. 664.

CHAPTER VII

THE COTTON FACTOR AND PLANTATION SUPPLY

THE origin of the cotton factorage system goes back to the early years of Virginia when the tobacco grower shipped his crop to England, entrusting it to a commission merchant who sold it and bought with the proceeds the various articles of merchandise ordered by the planter.[1] From Virginia the system spread throughout other parts of the South, for it was in cotton that the factorage system reached its greatest development, became most powerful, and flourished the longest. Cotton's nonperishable nature lent itself peculiarly to the concentration of the product in the seaport towns at a time when transportation and warehouse facilities were poor, for rough handling and long delays would have destroyed the value of other agricultural products.[2]

As the cotton industry grew, the relationship between planter and factor became more binding. There was little chance for the planter to escape, for if he sold it to some local purchaser it would have to be resold to the commission merchant in a port town. Thus, additional handling charges reduced the final profits, which made it more desirable for the planter to deal directly.[3] On rare occasions, planters attempted to deal directly with Liverpool, but the distance involved made the market even more uncertain than in New Orleans or Mobile. Nevertheless, the cotton factor did not main-

---

1 John Spencer Bassett, *The Plantation Overseer*, p. 221.

2 Alfred H. Stone, "The Cotton Factorage System", *American Historical Review*, XX, 559.

3 Anne Royal, *Letters from Alabama*, p. 162.

tain quite the same authority over his client as did the London merchant over the Virginia tobacco planter, partly because the cotton producer could keep informed as to price fluctuations and, if desirable, visit the city during the selling season.[4] Moreover, the Mobile factor did not in every case buy supplies from the proceeds of the sales of cotton. Orders might come to him during the spring for plantation supplies and not be paid for until the crop was sold in the fall, or local merchants might be authorized to draw against the account of a certain planter who had left a part of the proceeds of the preceding year on deposit with the factor. Again, it often happened that a comparatively large planter sold his cotton through two or three factors, and supplies bought through one might be paid for by drafts on another.

Between the years 1820 and 1830 no fewer than twenty-four commission merchants ran advertisements in the Mobile papers.[5] Among these were such names as J. F. Ross, Rives and Mather, Franklin Robinson, Jere Austill and other persons who were later to become closely identified with the cotton factorage business of Alabama. In addition, there were business notices of McGehee and Scott and Walter B. Lucas of Montgomery and Joseph Hutchinson of Augusta, Georgia.[6] Though the Panic of 1837 caused many factors to become insolvent, the years following brought many new men into the field, as well as an increased number of partnerships.[7] Even during normal times, commission merchants found it convenient to combine their businesses in the form of partnerships, and it was not unusual for a man who had

---

[4] Bassett, *The Plantation Overseer*, p. 224.

[5] See *Mobile Commercial Register*, 1820-1830.

[6] *Ibid.*, Nov. 22, 1823; Jan. 10, 1826.

[7] *Mobile Mercantile Advertiser*, Feb. 20, 1836; *Mobile City Directory*, 1837; *Mobile Commercial Register and Patriot*, June 8, 1838.

been in business for a number of years to have been associated at one time or another in several different firms.[8]

Even as early as 1819 factors were prone to make large advances secured by cotton. In that year John W. Walker borrowed over $9,000 from Thomas Percy and offered as collateral his crop and bank stock.[9] A year later J. S. Walker wrote his brother, John, that his cotton crop in Autauga County would be shipped to New York through factors in Mobile who had offered an average price of fifteen cents a pound.[10] Apparently, competition among commission merchants was keen from the very first, and every inducement was offered to urge the planters up the rivers to take advantage of the opportunities set forth in the advertisements. In this respect the following business notice is typical:

> Commission Business and Cotton Brokerage
>
> The Subscriber having taken the large and commodious store No. 2 Hallets Wharf, lately occupied by Messrs. F. W. and J. Armstrong, offers his services to his friends and the public as Commission Merchant and Cotton Broker, and hopes by attention and ability, to merit the patronage of all who may entrust their business to him.
>
> From the contiguity of his store to the water and wharf, Merchants and Planters in the interior will find their expenses lessened considerably by sending their cotton to the Subscriber either to ship or sell.
>
> A. F. Stone[11]

Charles Thomas, a former sugar factor from Louisiana who moved to Mobile in 1822, appeared to be engaged not only in the cotton commission business, but sold all kinds of goods, dealt in houses, lots and slaves, collected money, discounted notes at sixty or ninety days, and

---

[8] *Mobile Commercial Register*, Oct. 26, 1824.

[9] *John W. Walker Papers*, Nov., 1819 - Oct., 1820.

[10] *Ibid.*, April, 1821.

[11] *Mobile Commercial Register*, Dec. 24, 1821.

also acted as a ships' broker.[12] Gaines and Fowler, in an effort to secure more business, advertised a list of references consisting of C. Purdy and Company of New York, Thomas P. Cope and Son, Philadelphia, Keller and Foreman, Baltimore, Ben J. Harris, Richmond, Morgan and Stone, Huntsville, G. and G. Saltonstall, Tuscaloosa, George S. Gaines, Demopolis, and Thomas H. Herndon, Erie.[13] Another commission merchant offered free wharfage and storage in a fireproof warehouse, as well as liberal advances on all cotton which was handled by his firm.[14] Many newcomers in the factorage business were former planters or small-town merchants from the interior, who, foreseeing the opportunities in the business, abandoned their former occupations and came to Mobile.[15]

In 1824, the commission merchants of Mobile offered a premium for the best lot of cotton, not less than ten bales, which should be brought into market during a given period of the shipping season. This was done, no doubt, in the hope of encouraging Alabama planters to raise a better grade of cotton and to take more pains to see that it was properly ginned and packed. It very frequently happened that cotton of a poor grade was shipped to New Orleans, where the marks from the bales were erased, and exported as Mississippi or Louisiana cotton. For such frauds Mobile merchants were often held liable.[16]

Fraudulent methods in baling cotton brought a storm of protest from local exporters. One of these was the practice of packing good cotton on the outer side of the bale from which the samples were taken, while the interior was made up of a poorer grade. Sometimes stones,

---

[12] *Ibid.*, Feb. 28, 1822.
[13] *Ibid.*, Nov. 25, 1822.
[14] *Ibid.*, Feb. 27, 1822.
[15] *Ibid.*, Oct. 22; Nov. 16, 1824.
[16] *Ibid.*, May 18, 1824.

dirt or trash was placed in the middle of the bale in order to increase the weight, and several gallons of water were also found useful for the same dishonest purpose.[17] So widespread did these fraudulent methods become and so numerous were the rebates paid by commission merchants to exporters in New York and Liverpool that factors threatened to publish in the newspapers the name of every planter who indulged in such practices. On one shipment alone Messrs. Barclay and Company of Liverpool filed claims for a shortage of 3,867 pounds because of dishonest packing.[18] The state of affairs caused by this practice became so embarrassing that the cotton factors of Mobile petitioned the legislature for the passage of a law which would give more adequate protection. In 1818, such a law was passed, but it provided for a penalty only at the discretion of the court, in addition to the damages recovered at suit by the party aggrieved,[19] and in almost every case it was the commission merchant who suffered rather than the producer.[20]

However, deception and questionable practices were not all on the side of the planter. Factors were often accused of advising their clients to sell two or three days preceding a rise in price. Their locations in seaport towns gave them an advantage of receiving advanced news of market quotations as well as contemplated changes. By the time such information reached the interior the advantage was lost for the local producer. Not infrequently, on the prospect of a rising market, merchants bought their clients' cotton on their own accounts, which practice bordered on the pale of dishonesty.[21]

The costs of marketing cotton varied somewhat in

---

17 *Alabama Republican*, June 22, 1821; *Mobile Commercial Register*, Dec. 29, 1828.

18 *Ibid.*

19 Toulmin's *Digest*, p. 850.

20 *Mobile Commercial Register*, Dec. 29, 1828.

21 *The Tuscumbian*, June 27, 1825.

proportion to the distance hauled, the stage of the rivers, and the condition of the bales when they were received by the commission merchant. Practically the only standard item among the fixed charges was the 2½% commission for selling.[22] Freight from the Tennessee Valley, even on flat boats, down the Tennessee and Mississippi Rivers averaged about $3.50 a bale,[23] whereas Bolling Hall of Autauga County and Charles Tait of Wilcox paid only a dollar a bale for transportation to Mobile.[24] In 1837, the charges on 120 bales sold by Manusel White of New Orleans for A. J. Hutchings of Lauderdale County amounted to $595.10, which was distributed as follows: freight, $266.76; drayage, labor and weighing, $36.00; storage, $24.00; fire and river insurance, $84.00; and commissions at 2½%, $184.34.[25] With cotton bringing 6¼ cents a pound in 1842, Israel Pickens sold 100 bales at a gross profit of $2,837.50 but was forced to pay $246.13 for marketing costs.[26]

If planters shipped from public landings along the rivers, additional charges had to be paid. For example, one producer had to pay $63.75 for the privilege of loading 170 bales from the municipal wharf at Montgomery.[27] During a prolonged drought in 1855 the upper part of the Tombigbee River was navigable for only small light-draft boats, and so great was the demand to transport cotton to Mobile that William P. Gould was forced to pay four dollars a bale freight for a shipment of forty bales on June 25. In addition, a charge of twenty cents per bale was made for loading at Miller's

---

[22] This percentage remained the same from the early days of statehood until 1900.

[23] Bills of sale by Reynolds and Rapier, April, 1820, and Brown and McAdam, Dec., 1822, *John Coffee Papers.*

[24] *Bolling Hall MSS.*, May, 1834; *Charles Tait Papers,* Oct., 1824.

[25] *John Coffee Papers,* Feb. 23, 1837.

[26] *Israel Pickens Papers,* Dec., 1842.

[27] *Thomas Brown Journal,* Oct., 1845.

Landing, and this added to the other costs made the charges so high that Gould decided to take out no river insurance and paid only the usual charge of eleven cents for fire insurance. Even though his cotton sold for 11½ cents, over one-tenth of the proceeds was taken up by transportation and selling costs alone. This, however, must be noted as an exceptional case brought about by the unusually low stages of the river.[28]

Whenever cotton was sold, the commission merchant usually sent the owner a bill of sale which indicated the number of bales in the transaction converted into pounds, the price per pound that the particular shipment brought on the market, less an itemized list of the charges, which when deducted from the gross sales indicated the net profit to the owner. In certain cases, however, a factorage report was submitted which was actually a statement from the commission merchant's books indicating the standing of a particular planter at a certain date. Thc following is a typical example:

Account of Samuel Pickens with Franklin Robinson
July 1, 1832

	Date			Interest	Principal
1830					
	June	29	Cash paid for candles	$1.92	$24.85
	Nov.	28	Cash lent Wm. Daniel	.40	10.00
1831					
	Jan.	11	Bill for rope	1.04	28.91
	Mar.	1	Bill for potatoes		2.70
	Apr.	1	(93 days discount on U. S. Bank note)	.94	47.00
		26	Postage for letter		.19
	May	5	Cash forwarded by M. Hobson		1000.00
		12	Commission on 41 bales		
			Balance July 1		$1819.10[29]

[28] *William P. Gould Papers.*

[29] *Samuel Pickens Papers.* The balance on July 1 appears to be inconsistent with the other figures as stated. No amount was given for "commissions on 41 bales".

As a usual thing planters seemed to have had the utmost faith in the honesty of the merchants through whom they dealt. Frequently, however, misunderstandings arose in which each party endeavored to convince the other of his mistake. In December, 1834, a cotton factor in Mobile by the name of Sheppard was accused by Israel Pickens of failing to sell his cotton when ordered to do so and of overcharging him for supplies purchased. Sheppard denied the charge and asked Pickens to correct it for fear of its injuring his business connections in the Black Belt.[30] Once cotton had come under the control of the factor it was usually considered a better policy to leave the actual marketing in the hands of the agent. "We do believe it to be to the interest of the Planter to leave the management of his cotton to his merchant after it arrives here," wrote Tait and Company to Samuel Pickens, who insisted that a certain number of bales be removed from Casay's warehouse.[31]

A discrepancy of 4,000 pounds in a shipment of cotton to New Orleans caused Mrs. Mary Coffee in 1841 to accuse Manusel White of dishonest methods.[32] A number of bales had been damaged by water, it was afterwards proved, but while the insurance adjustment was being completed the factor wrote Mrs. Coffee as follows:

> Common courtesy tells us to treat your communication of the 13th inst. with respect, yet the writer of the same deserves no consideration from us, and we are somewhat surprised that after transacting the business of your husband and self for so many years, that at this late date the writer should insinuate that we had willingly wronged you out of a cent.[33]

The services rendered by commission merchants in nearly every case were the same. Each one charged

---

30 *Israel Pickens Papers.*

31 *Samuel Pickens Papers*, Dec., 1844.

32 *John Coffee Papers*, April 13, 1841.

33 Manusel White to Mary Coffee, April 23, 1841, *John Coffee Papers.*

2½% for selling cotton, bought supplies for planters and made advances secured by the next year's crop. The planter's choice then must have been directed very largely by a feeling of friendship or personal association. In some instances family influences were an aid in forming business connections.[34] In Wilcox, Dallas, Marengo and Greene Counties, Rives, Battle and Company, Boykin and McRae, James Crawford, Malone and Foote, Hartwell W. Davis and Alexander Pope and Son did quite an extensive business, while in the Tennessee Valley, where nearly all the factorage connections were with New Orleans, Manusel White, Fearn, Donegan and Company, Reynolds and Rapier, and Brown and McAdam seem to have been the most popular.[35]

The firm of Boykin and McRae began as Norris and Boykin, but in 1842 Norris retired and Colin McRae of Pascagoula, Mississippi, was taken into partnership.[36] Boykin had come originally from Dallas County, where both he and McRae had large family connections. During a part of the winter McRae solicited business in Clarke and Wilcox Counties in order to increase the well-established clientele already shipping cotton through Norris and Boykin.[37] By 1852 the chief clerk of Boykin and McRae had become a partner, and the firm henceforth was known as Boykin, McRae and Foster.[38]

It was not uncommon for the owner of a general merchandise business to buy cotton and finance planters on a small scale. For example, Joseph W. Wilson of Montgomery, who was forced into bankruptcy during the Panic of 1837, began again as a commission merchant

[34] Among the clients of Boykin and McRae were many kinsmen of Dallas County.

[35] This estimate is based not only on bills of sale but also on newspaper advertisements in the various counties.

[36] Colin J. McRae to Mrs. Elizabeth McRae, Sept. 20, 1842, *Colin McRae Papers*.

[37] Same to same, *ibid.*

[38] *L. M. Burford MSS.*

and nurseryman. On certain occasions, cotton was bought outright for purposes of speculation, and some, no doubt, was taken in payment for fruit trees which were purchased by Black Belt planters. However, there are many entries on Wilson's books which clearly indicate that cash was lent on current crops and that mortgages were held. From March to June of 1840, Wilson's books show that the gross from his commission business was $14,595, and that from October, 1840, to July, 1841, cotton transactions amounted to $24,921.39.[39] It is also significant to note that his merchandise account during the first six months of 1841 was slightly less than $2,000.[40]

The commission house of Alexander Pope and Son represents one of the more successful large businesses in Mobile from 1829 to 1847. In October, 1829, Pope, who had previously been a clerk in the Cahaba land office, moved to Mobile to engage in the cotton commission business.[41] Shortly after arriving he wrote Charles Tait at Black's Bend to notify his friends in the neighborhood of Claiborne that he would be pleased to handle their cotton.[42] Aided by his connection with George Hold and Company, cotton brokers of Liverpool, and his own conservative nature, Pope was able to weather the financial turbulence caused by the Panic of 1837.[43] From Liverpool he wrote his son in 1839:

> People are running mad about cotton and prices have exceeded all bounds, but there will be a day of reckoning by those who deal too deeply. Take my advice and don't touch it. Sell all you can but in no case buy any cotton. There are first rate men here who are willing to make an advance of from 7 to 8 cents per lb. on fair cotton.
>
> 15000 bales sold here on speculation. Prices have ad-

---

39 *Ledger and Account Books of Joseph W. Wilson,* 1840-1841.

40 *Ledger of Joseph W. Wilson,* June, 1841.

41 *Charles Tait Papers,* 1829.

42 Pope to Tait, Oct. 10, 1829, *ibid.*

43 *Alexander Pope MSS.,* 1837.

> vanced, but when men are mad there is no telling what they will do. My object in writing is to warn you of the danger and advise you to keep out of it. You might make a large profit, but as in other forms of gambling, you have about four chances to one against you, and you might involve yourself in difficulties which would ruin you forever. Sell, therefore, all you can, but buy none and ship none, but I presume few planters would be willing to ship without drawing largely, and it is against this that I warn you . . . .[44]

The primary purpose of Pope's visit to Europe in 1839 was to collect claims for a partially destroyed cargo of cotton shipped on the schooner *Horacio*.[45] The underwriters apparently acted very cleverly and offered to settle for 900£ with an assurance that the remainder would be paid when the part of the cargo which was saved could be sold. Pope, however, deemed it wiser to refuse payment and await the outcome.[46]

In 1843 cotton to the amount of 2,744£ was sold by Pope to Isaac Low of Liverpool,[47] and the next year 4,520£ for commissions, interest, expenses for handling cotton, and collections was owed by Pope to this English firm.[48] During the year 1843 the financial statement of Alexander Pope and Son showed a gross business of $220,234.75, with a net profit of $67,291.84. The profit and loss statement was distributed as follows:

Expense	$ 2817.49	Balance	$ 53943.72
Exchange	212.63	Interest	1998.98
Merchandise	897.93	Insurance	668.42
Balance to credit	67,291.84	Commission	14608.77
	$71,219.89		$71,219.89[49]

[44] Alexander Pope to A. L. Pope, Jan. 1, 1839, *ibid.*

[45] Same to same, Jan. 20, 1839, *ibid.*

[46] Same to same, *ibid.*

[47] *Alexander Pope MSS.*, 1843.

[48] *Ibid.*, 1844.

[49] Profit and Loss Statement, 1843, *ibid.*

The net proceeds on a 921 bale shipment in 1846 amounted to 6,629£ and involved an expenditure of 841£, of which 724£ went for storage and fire insurance, 23£ for receiving and delivering in England, 18£ for bank commissions, 37£ as brokerage fees for purchasing and the same amount as brokerage fees for selling. As a usual thing, foreign sales were conducted through a cotton exporting company, but on this occasion transactions were handled directly with the individual purchasers. John Acton and Company bought 200 bales, Wild and Wright, fifty, Fairclough and Chinn, fifty, Joseph Butterworth and Son, 300, H. Castellian and Sons, fifty-one, H. H. Stott, seventy, and Swainson, Birley and Company, 200 bales.[50]

Commission merchants lent out to planters a great deal of their surplus capital on notes and bills receivable. Between February and May of 1844, Pope placed thirty-two notes running from $100 to $3,276 in the hands of Newton St. John for collection. The average amount of each note was from $1,200 to $1,500 and the total sum was $33,384.60.[51] Some ran for as short a time as three months, but the great majority were to mature in eight or nine months, which was the time from the sale of one crop to another. Listed in the "bills receivable ledger" were not only the familiar names of many Black Belt planters but also of prominent Mobile business men.[52] The factor, then, maintained with the planter the joint position of commission merchant and banker, for he not only sold his cotton but also credited the proceeds to the planter's account.[53] Factors were sometimes tempted to act in the double role of both seller and buyer, especially if the market gave indications of a sudden

---

50 *Alexander Pope MSS.*, 1846.

51 *Ibid.*, 1843.

52 *Ibid.*, 1844.

53 Lewis C. Gray, *History of Agriculture in the South to 1860*, II, 711.

rise, but it was against this type of speculation that Alexander Pope so emphatically warned his son in 1839.[54]

Pope died in 1847 and left an estate of $118,038.04 which consisted of land, bank stock, rent on his warehouses and uncollected notes. His son, A. L. Pope, then formed a connection with Powers and Company, commission merchants of New Orleans.[55]

In nearly all cases, cotton was shipped on consignment to factors in the seaport towns. The price was controlled largely by the grade of the cotton, the season of the year, and the future prospects of the market. In 1857, L. McQueen wrote R. L. Jenkins: "The cotton market having recovered slightly from its depressed condition, and seeing little or no prospect of any improvement soon, I have concluded that it is to your interest to sell."[56] The buying interests came to be carried on largely by brokers who acted on behalf of foreign dealers or manufacturers, though it was quite common for British and French spinners to have their personal representatives in Mobile who were known as cotton buyers.[57] The buyer, usually authorized to purchase a certain number of bales of a particular grade of cotton, placed his order with a broker, who in turn dealt with the commission merchants who filled the order from their warehouse stocks.

Cotton en route to market was insured at the owner's expense, and in most cases both fire and river insurance were carried. The Augusta Fire Insurance Company offered rates of one-half of one percent on any good steamboat or barge and seems to have had a rather extensive business among Black Belt planters.[58] Factors themselves carried coverage on all cotton placed in their

---

[54] See pp. 150-151 above.

[55] *Alexander Pope MSS.*, 1847.

[56] *Burford MSS.*, Dec. 19, 1857.

[57] Harleston Broun was one of the most successful cotton buyers in Mobile, representing French mill interests.

[58] *Charles Tait Papers*, Dec. 27, 1832.

charge, for considering the frequent damage from fires and freshets in the rivers it was almost a necessity. Manusel White wrote General Coffee that "your cotton has been insured for $45 per bale and as I have an open policy for 2,000 bales, they will be under insurance the moment they arrive".[59]

The nonslaveholder and small planter frequently sold their cotton through a local merchant who was willing to make small advances on crops. The country storekeeper then became a sort of intermediate factor and sold cotton through commission merchants in the port cities as did the large planter. J. M. Bradford of Wetumpka owned a warehouse in connection with his general store, and a close study of the cotton receipt books indicates that the average-sized account was for two or three bales. In 1849, the largest account was for ten bales, and there were several of only one bale each. Many of these were undoubtedly the accounts of nonslaveholding farmers who hauled their cotton from the adjacent hill counties to Wetumpka, which was at the head of navigation on the Coosa River. Factors in Mobile apparently paid Bradford a commission on all cotton he shipped to them.[60] Another merchant of Wetumpka in 1857 did not buy cotton outright but shipped it to Mobile on a consignment basis. However, he agreed to bear the wharfage and shipping charges, which were subtracted from his commission for selling.[61] At Whiteplains, in Calhoun County, A. A. Causby made advances to the small farmers and agreed to buy their cotton in lots as small as twenty pounds.[62]

Each river town had its intermediate factor who not only sold cotton for the small producer but also took orders for bagging, bale rope and other plantation sup-

---

59 *John Coffee Papers*, Feb. 24, 1832.

60 *J. M. Bradford Warehouse Receipt Book*, 1849.

61 *Account Book of C. J. W.*, Wetumpka, 1857.

62 *Account Book of A. A. Causby*, 1839.

plies.[63] A merchant at Erie on the Tombigbee River advertised once a week for a month that he advanced cash on cotton. "The subscriber will advance liberally on cotton which is designated for shipment from this and adjacent landings — John C. Phares, Erie, Alabama."[64] At Huntsville, in the Tennessee Valley, J. B. Peabody did an average monthly cotton business of from $1,200 to $1,500 between 1835 and 1837 shipping to factors in New Orleans.[65] General Coffee on one occasion sold twenty-four bales through C. McLemoore in Nashville, from which place it was transhipped to New Orleans.[66]

In contrast to the regular commission business, John Bowie opened in Mobile a cotton brokerage house which proposed to sell cotton for a straight charge of fifty cents a bale. This was entirely different from the methods of the regular factor who bought supplies on order, made cash advances and thereby took numerous monetary risks, but always at a 2½% commission for both purchases and sales. Bowie seemed particularly interested in devoting himself strictly to sales without incurring any of the other responsibilities of the regular commission business, and hoped that planters would find his plan satisfactory.[67] It is quite likely that the Panic of 1837, which brought financial ruin to so many commission merchants, may have suggested the idea, but there is no indication whether or not the plan was ever subscribed to. Credit was essential to the operation of the plantation system, and it was the factor himself who benefited not only from the commissions for purchases but also from the accrued interest as well.

As a rule, commission merchants discounted their paper with the Mobile branch of the State Bank, and

---

63 *Greene County Gazette*, Aug. 30, 1830.

64 *Ibid.*, Dec., 1831.

65 *J. B. Peabody Account Book*, 1835-1838.

66 *John Coffee Papers*, Dec., 1824.

67 *Selma Free Press*, Oct. 27, 1838.

upon examination of the Discount Books one can find the name of nearly every merchant.[68] Frequently during the thirties money was extremely scarce, and factors had trouble in negotiating paper.[69] On August 17, 1831, Pickens' factor, Franklin Robinson, wrote: "I am unable to raise more than $100 as there are no U. S. Bank notes in Mobile. However, on September 1st I hope to be able to sell exchange to the State Bank."[70]

So acute did the monetary situation become that in 1840 there was a large meeting of planters and factors in Mobile to discuss the extremely hard times and to consult as to the best means of relief in consequence of the low price of cotton and the scarcity of money. Resolutions were adopted that the Bank of Alabama in Mobile should advance a considerable sum "upon bills of exchange drawn by planters against shipments of their cotton, which will mature if drawn on any foreign place in six months from the date of shipment."[71] A committee was appointed to confer with the directors of the bank, who, after considerable deliberation, refused to give a direct answer, and it was generally inferred that they were unwilling to grant assistance in the manner solicited by the committee.[72] Another question which arose was whether cotton in the field could be used to pay current debts. The directors of the Montgomery Branch Bank decided that even October was too early for crops to be considered as assets.[73]

The relations between the cotton factor and the planter were of the most intimate and confidential character. The ties between them frequently were lifelong, and their

---

68 *Discount Book of the Mobile Branch Bank*, 1835-1841.

69 Alexander Lopez to Samuel Pickens, Nov. 27, 1833, *Samuel Pickens Papers*.

70 Robinson to Pickens, Aug. 17, 1831, *ibid.*

71 *Selma Free Press*, Feb. 29, 1840.

72 *Ibid.*, March 7, 1840.

73 *Minute Book of the Montgomery Branch Bank*, 1839-1843.

relations were of a social and personal as well as a business nature. Just how far this close personal association affected plantation policy in Alabama it is impossible to say, yet we are certain that the advice of the factor was frequently reflected in the planter's affairs. Thousands of dollars were advanced, in many cases merely upon the personal word of the planter with no formal security and only a memorandum to bear witness to the amounts involved.[74]

As stated before, it was usually left entirely to the judgment of the factor to choose the proper stage of the market to sell. White wrote General Coffee that "Tennessee Valley cotton is quoted at 17 cents, but I took the liberty to hold your bales hoping to get a better price".[75] Presents for the planter's wife were other indications of the close ties which existed between the producer and his commission merchant. In April, 1825, a large box of shrubs containing geraniums, cape jasmine and crape-myrtle was received at Florence by Mrs. John Coffee as a gift from Manusel White of New Orleans.[76] A small planter of Clarke County even wrote his factor in Mobile to tell him of the death of his wife: "After what is due to friendship I rite [sic] you these fue [sic] lines to inform you of the death of my wife — she departed this life on Wednesday mornin [sic] the foreteenth [sic] of this present month in great peace of congestive chills . . . which is the only reason of my riting [sic] these few lines."[77]

There is little evidence in Alabama to show that a single factor held absolute control over the planter's crop, as seems to be the current belief.[78] Large

---

[74] Stone, "The Cotton Factorage System", *American Historical Review*, XX, 559.

[75] *John Coffee Papers*, Feb. 26, 1825.

[76] *Ibid.*, April 21, 1825.

[77] *Montgomery Advertiser*, April 14, 1855.

[78] See Alfred H. Stone, *American Historical Review*, XX, 563.

producers appear to have done business simultaneously with two or three commission merchants, and it was not unusual for a planter to draw on one to settle his account with another.[79] The common procedure in paying current accounts was to give drafts on the factor.[80] Tait gave an order on Boykin, McRae and Foster of Mobile in payment for two mules he purchased from a Kentucky drover.[81] In 1850, William P. Gould paid a local dry goods account with W. B. Johnson amounting to $68.63 by a draft on Rives, Battle and Company.[82]

The factorage business was not, however, without its hazards and burdens. Large-scale planting enterprises required not only considerable credit facilities but heavy labor equipment; and unfortunately, there was in such enterprises something of both the nature and spirit of games of chance. With the planter it seemed to be either a feast or a famine. According to the very nature of the system, this chance in some measure had to be shared by the factor, who was forced to develop a scale of charges which were correspondingly high. A depressed cotton market often found the planter struggling under a burden of debt which steadily increased, and after a few years it was natural for him to feel that he was hopelessly harnessed to the factor's plow. On the other hand, there were enough abuses of confidence and credit on the planter's part, such as extravagant living on funds furnished purely for business purposes, or neglect of their common interests with resulting heavy losses to both, to make the factor feel that though his profits in good times were large, they would still be insufficient to balance the risk he ran.[83]

---

[79] On April 22, 1845, Tait drew on Boykin and McRae to pay a note of $1,000 held by Rives, Battle and Company. *James A. Tait Plantation Memorandum Book.*

[80] *Samuel Pickens Papers,* Oct. 6, 1831.

[81] *James A. Tait Plantation Memorandum Book,* 1851.

[82] *William P. Gould Papers,* 1851.

[83] Stone, *American Historical Review,* XX, 560.

Factors were frequently forced to spend considerable sums in entertaining their clients, who came frequently to Mobile on business trips. A single bar-room account of Colin McRae is said to have run over $1,700 in one year, and the greater part of this amount went for the entertainment of Black Belt planters with whom he had or expected to have business connections. Franklin Robinson wrote Israel Pickens that heavy advances to planters in 1834 prevented him from paying his current obligations. However, he felt that the money was still good even though it was slow to come in, and if his friends would coöperate he could still continue to do business.[84]

Transportation was another problem which affected the prosperity of the factor in Alabama. Prolonged droughts frequently prevented boats from reaching the upper parts of the rivers, and sometimes cotton was delayed many weeks in reaching market. Such was the case in January, 1854, when J. C. Lewis of Mobile wrote: "We have had no rain here yet and the rivers are too low to let cotton come down and our factors are in a mighty bad fix. They pay everything however as yet, but things look gloomy as there is no prospect of rain."[85]

In 1855, William P. Gould borrowed $5,489.78 for thirty days on a note held by M. N. Smith of Mobile. When the time expired and the note was still unpaid, the factor wrote as follows:

> At the time I loaned you the money I was very much pressed for cash and would not have loaned that amount at 20 percent for the entire season. These extraordinary times of low rivers have caused an oppressive demand for money.
>
> I understand you sold your Florida lands to Croom and request that you use some extraordinary exertion to get that note cashed. Croom, himself, might cash it at a sacrifice, and whatever sacrifice you pay would not be

[84] *Israel Pickens Papers.*

[85] J. C. Lewis to Jones Fuller, Jan. 18, 1854, *Fuller MSS.*

> as great as perhaps the forfiture [sic] of your word may yet cost me. I have tried to negotiate your note but can't.[86]

Probably one tendency of the factorage system in Alabama, especially until the coming of the railroads in the fifties, was to retard the growth of urban communities in the interior of the State. Mobile, the factorage center for the southern and central parts of the State, and New Orleans for the Tennessee Valley, were supply depots from which the planter bought his plantation necessities in bulk instead of through the local merchant. This, no doubt, was partially responsible for the great concentration of capital in the seaport cities.

As early as 1805 Mobile was engaging in a thriving trade with settlements in the interior. The great trading house of Panton, Leslie and Company of Pensacola, famous for its Indian trade throughout the entire Creek country, had taken into partnership a Mr. Forbes of Mobile, where Panton, Leslie, Forbes and Company had opened a branch operated under the name of John Forbes and Company.[87] Reports and manifests between 1805 and 1811 indicate that Forbes was importing, principally from New Orleans, cloth, salt, blankets, thread, spelling books, guns and cotton bagging, as well as considerable quantities of food supplies.[88] By 1816, imports at Mobile from New York, Boston and New Orleans were estimated at a million dollars, and there is evidence to show that the greater portion of this was transported up the rivers to planting communities within the trade area.[89] Bacon sold for sixteen cents a pound, sugar for fourteen cents, coffee for twenty-nine cents, flour for fourteen dollars

---

[86] *William P. Gould Papers*, Feb., 1855.

[87] *Ellicott's Journal*, p. 211; Richard L. Campbell, *Historical Sketches of Colonial Florida*, p. 100.

[88] *Reports and Manifests*, Mobile, 1805-1811.

[89] Warden, *Statistical, Political and Historical Account of the United States of America*, III, 38.

a barrel, gingham at sixty cents per yard and salt for $1.25 a bushel.[90] In 1816, the general store of Bidgood and Walton at Forkland on the Tombigbee River bought $2,498.85 worth of merchandise consisting of needles, handsaws, spectacles, blankets, cotton goods, tobacco and books from Cumming and Ramsay of New Orleans through an importing house in Mobile.[91]

There seemed to be a rather marked scarcity of trading centers, for even the river towns found it difficult to compete with the well-established commission houses in Mobile.[92] Wetumpka, however, at the head of navigation on the Coosa River, was an exception, for in 1843 a traveler noted that he found produce cheaper there than in other parts of the State because of the great volume of trade and the competition with Montgomery.[93] Even in Montgomery, Robertson maintained that no firm of merchants was ever known to have become particularly rich. The small planters and farmers had only small accounts, while the large planters bought their negroes in Virginia, their mules in Kentucky and Tennessee and their plantation supplies in Mobile, where factorage connections were maintained.[94] This situation also retarded the opportunity of Montgomery merchants to speculate in cotton on any large scale.

The cotton factor not only sold the planter's cotton but also bought his plantation supplies at the accustomed commission of 2½%. It sometimes happened that accounts were opened at various grocery, negro supply and drygoods concerns where the planter did his own purchasing but paid his bills by drafts on the factor. On rare occasions, settlement might even be made by means

---

90 *Hugh Rolston MSS.*

91 *Invoice Account Book,* 1816.

92 *DuBose Canebrake MSS.*

93 *Dubois Diary,* March 6, 1843.

94 W. G. Robertson, *Recollections of Montgomery County,* pp. 11-12.

of a note held for collection by the branch bank. For example, on December 28, 1839, James Tait purchased drygoods from Norris, Crane and Keith amounting to $788.97 which was to be paid by a note to become due on February 1, 1840.[95] The Taits apparently had open accounts with R. L. Watkins and Company, dealers in drugs and paints, M. I. Stetson, dealer in boots, shoes, hats and clothing, DeForest, Morris and Wilkins, hardware and cutlery merchants, Ross and Marshall, grocers, Greene and Robbins, saddlery and harness dealers, P. Cullen Drygoods Company, Kelley and Allen, dealers in furniture, R. R. Belknap, hardware merchant, and S. G. Kellogg, bookseller and stationer.[96] In 1832, lumber for negro houses on the Tait place was purchased from a local saw mill at an average price of eight dollars a thousand feet, though nails, hinges and other items of building hardware were bought in Mobile.[97]

Among the interesting items purchased by the Taits was a cotton gin sent from Charleston for a month's trial at the manufacturer's expense and guaranteed to give perfect satisfaction in operation. Another was a coffin Charles purchased in the early spring of 1834, only to have it delivered with due apologies by B. Newhouse and Company in October of that year.[98] It is reasonable to suppose that the deceased must have been interred months before in a rough pine box fashioned by the plantation carpenter. During one month in 1840 James Tait's purchases amounted to over $900. The greater part of this sum was spent for staple groceries such as molasses, bacon, flour, sugar, lard, salt, coffee, pork, lemon syrup, ham, whiskey, wine and brandy, while the other went for carpenters' tools and sundry forms of hardware.[99]

---

[95] *Charles Tait Papers.*

[96] *Ibid.,* bills rendered between 1830 and 1840.

[97] *Charles Tait Journal,* 1832.

[98] *Ibid.,* 1834.

[99] *Ibid.,* May 16, 1840.

Payment in cash even for locally purchased items seems to have been rare. L. M. Burford purchased from his neighbors a plow, a log chain, a spinning wheel, a large iron pot, three axes, several head of cattle, two yoke of steers and a wagon, all amounting to $132.53. Not a single cash transaction took place among the lot, but each item was paid for by an individual note.[100] Pickens was forced to buy on credit 3,000 pounds of bacon at eight cents a pound when a cash settlement would have brought him a better price.[101]

Even though bagging, rope, pork, sugar, flour, coffee and molasses apparently were the most frequent plantation purchases, sundries of various kinds were nearly always included. Ice in the summer and oysters in the winter were by no means unknown.[102] Negro hats and shoes were usually purchased in comparatively large quantities, while calico was bought by the bolt and made into clothes on the plantation.[103] During the fifties the local merchant began to sell a small percentage of incidental items, but never made more than slight inroads into the monopoly of the commission merchants and wholesale houses in Mobile.[104] The warehouse ledger of C. Hall and Company at Portland, in Dallas County, indicates that by far the greatest amount of supplies going to planters in the vicinity of Tilden and Molett's Bend came from Mobile, and it can be said that this was typical of other parts of the Black Belt, especially where estates had access to river traffic. Accounts on Hall's books ranged from ten dollars to $500, with the average purchase amounting to probably fifty dollars.[105]

---

100 *Burford MSS.*, Dec. 16, 1844.

101 *Samuel Pickens Papers*, Aug. 6, 1845.

102 *William P. Gould Papers*, Diary, 1853.

103 *Burford MSS.*, Oct., 1855; *William P. Gould Papers*, April, 1859.

104 *Ibid.*, 1859-1860.

105 *Warehouse Ledger of C. Hall and Company*, 1859-1860.

The Tennessee Valley, with factorage connections in New Orleans, bought most of its supplies there, although before 1830 General Coffee bought considerable foodstuff from Dabney Morris and Josiah Nichol of Nashville. Flour from Kentucky and Ohio and pork from Tennessee were often purchased directly by planters in the neighborhood of Florence and Huntsville.[106] Freight rates for the 1,500-mile river trip from New Orleans in every case ran up the cost. In 1837 Coffee was forced to pay five dollars for freight, commission and handling charges on a twenty-dollar purchase.[107] On the eve of the Panic of 1837 commodity prices rose higher in the Tennessee Valley than in any other part of the State. Poor-grade coffee sold for twenty cents a pound, corn at two dollars a bushel, pork at ten cents a pound, eggs at thirty-seven cents a dozen, brown sugar for 12½ cents a pound, negro shoes for $2.42 a pair and small water buckets for a dollar each.[108] This situation, coupled with speculations in public lands, caused many planters to lose their entire holdings.

The Clays of Huntsville preferred to buy their furniture in Charleston or New Orleans. In 1844, Virginia Clay was instructed to choose suitable parlor furniture while on a trip to New Orleans and to pay for it by a draft on Fearn, Donegan and Company. The purchase was to be sent by river to Tuscumbia and then hauled overland to Huntsville.[109] Again, in 1850, C. C. Clay, Jr., while on a business trip, was asked by his mother to make purchases of china and furniture in Charleston. She also requested that he select for her some parlor draperies, and suggested that he look around in both New York and Philadelphia, where the styles would be more up to date. C. C. Clay, Sr., authorized payment

---

106 *John Coffee Papers,* 1817, 1822, 1825.

107 *Ibid.,* Jan., 1837.

108 *J. B. Peabody Account Book,* March, 1837.

109 C. C. Clay, Jr., to Virginia Clay, March 20, 1844, *Clay MSS.*

by a draft on Bradford's house for the necessary amount.[110]

The upper Coosa Valley had its trade connections almost exclusively with Rome. By 1850 the Georgia railroads formed a connecting link with Charleston and Savannah, from which ports cotton could be shipped abroad and plantation supplies brought into the interior. The upper stretches of the river were navigable for small boats as far up as Rome. B. C. Yancey, who owned a comparatively large plantation in Cherokee County, not only had factorage connections, but also bought all foodstuffs and negro supplies in Georgia.[111]

Between the years of 1855 and 1860 the price of provisions rose to a peak all out of proportion to the price of land and cotton. At Huntsville Clay complained that the cost of living had reached a point far in excess of the diminishing valuation of his property.[112] In the Black Belt, planters complained about the outrageous price of pork, which during the year 1857 remained between eight and nine cents a pound. The *Alabama Beacon* suggested that purchasers refuse to buy at such prices and advocated that they investigate pickled pork in New Orleans, which retailed for twelve dollars a barrel.[113] Commodity prices in Mobile were scaled up in the same proportion. Bagging sold for twenty cents a yard, bacon at fifteen cents a pound, cheese at sixteen cents, flour at ten cents, lard at sixteen cents, sugar at ten cents a pound and molasses at forty cents a gallon.[114]

During the same period there began a rather noticeable effort on the part of the railroads to compete with

---

110 H. L. Clay to C. C. Clay, Jr., June 5, 1850, *ibid.*

111 *B. C. Yancey Coosa River Plantation Account Book*, 1855-1860.

112 H. L. Clay to Virginia Clay, Feb. 5, 1855, *Clay MSS.*

113 *Alabama Beacon*, Dec. 18, 1857.

114 *Clarke County Democrat*, Feb. 14, 1856; *Warehouse Ledger of C. Hall and Company*, 1858-1860.

river traffic. The chief engineer of the North and South Railroad stated in his report to the board of directors that bacon and flour could be hauled from Kentucky to Montgomery, Selma or Tuscaloosa one cent per pound cheaper than would be the case if it were brought down the Mississippi River to New Orleans, transhipped to Mobile and thence sent up the rivers. There is some evidence that these reduced freight rates came about because of the fear that the counties of Benton, St. Clair, Cherokee, Blount, Jefferson and other upper counties of the State might be developed into a wheat-raising section, which would cut to a minimum the longer haul from the West. "Like the neighboring counties of Georgia and Tennessee", the report continued, "when once they get a market for their grain they will bid farewell to cotton as a staple. Northwest Georgians today make more money on wheat alone than any agricultural people in the United States."[115]

As has been previously stated, the chief function of the community store before 1850 was to supply the small farmer and nonslaveholder. During the ten years prior to the Civil War the business of the inland merchant, especially in the river towns, had reached the point where his stock of goods and prices were more in line with those of the larger houses in the port cities.[116] In the small slaveholding counties of Alabama, the country merchant was willing to take in small amounts of cotton, as well as other types of agricultural produce, and make advances on that basis. In Jefferson County, even as late as 1855, Peter Downey's store was engaging in a veritable system of barter in which he swapped sugar and coffee, rope, bagging and other staples for cotton, chickens and eggs.[117] In addition, a blacksmith shop was

---

[115] *Report of the Chief Engineer to the President and Board of Directors of the North and South Railroad Company*, 1855, p. 15.

[116] See *William P. Gould Papers* and *Burford MSS.*, 1850-1860.

[117] *Account Book of Peter Downey*, 1855.

often run in connection with the crossroads store, and it was here that planters usually sent their horses to be shod.[118]

In the early days, the prices of country merchants were naturally much higher than corresponding prices in Mobile or New Orleans. A typical store in the Canton community in Wilcox County apparently specialized in the sale of liquor. Its books show that a large percentage of the sales were purchases of whiskey by neighboring planters. One man averaged more than fourteen gallons of whiskey, besides large amounts of beer and cider, monthly for nearly a year.[119]

If importations of foodstuffs through Mobile are any indication of the growth of the plantation system in Alabama, rapid strides were made during the decade preceding the Civil War. The following is a partial list of importations of staples in 1850 and 1860:[120]

	1850		1860
Flour	67,000 Bbls.	Flour	140,691 Bbls.
Corn	24,314 Sacks	Corn	31,619 Sacks
Lard	10,582 Kegs	Lard	24,614 Kegs
Pork	17,517 Bbls.	Pork	31,092 Bbls.

It is logical to conclude that the greater portion of the commodities listed above found their way to planters in the interior of the State, since Mobile in 1860 had a population of only slightly over 30,000. Such vast quantities of pork, lard and corn might further be taken as an indication that planters were buying on an open market at high prices foodstuffs and feed for stock which they should have been raising for their own use.

Alabama also enjoyed a rather lively grain trade

118 In 1856 Burford paid $45.51 to a local blacksmith for the shoeing of twenty horses and minor repairs on plows. *L. M. Burford MSS.*

119 *Ledger and Day Book of the Canton Community Store*, 1829-1830.

120 *DeBow's Review*, Dec., 1850, p. 660; Nov., 1860, p. 665.

with Kentucky and Ohio, as flour was usually designated as coming from one or the other of those states.[121] Raising mules for a Southern market was also an important industry in both Tennessee and Kentucky, since the Southern planter believed it to be more profitable to buy his draft animals and concentrate his energies on cotton.[122] Sometimes pork slaughtered in Kentucky was sold from "pork wagons" which traveled from place to place and traded with the plantation owners. Hog drovers brought their herds through the country, as well as did dealers in cattle and sheep.[123] Thus, while the planter was concentrating on his one staple money crop he was forced to buy from the neighboring small farmer and the Kentucky drover a great portion of his plantation food supply.

---

121 *The American Cotton Planter*, I, 11.

122 T. D. Clark, "Livestock Trade between Kentucky and the South, 1840-1860", *Register of the Kentucky State Historical Society*, XXVII, 573.

123 *Ibid.*, p. 575; William E. Dodd, *The Cotton Kingdom*, p. 401.

CHAPTER VIII

PROFITS IN PLANTING

THE primary purpose of the Greensboro Agricultural Society, in the words of its President, Robert Withers, was for the "preservation and amelioration of the soil on which we entirely depend. We have not availed ourselves of the many lights reflected on agriculture by researches of modern times."[1] Such a statement coming in 1830, scarcely eleven years after Alabama had been admitted to statehood, might cause one to believe that the problem of soil exhaustion was fresh in the memories of those who had recently come from the older states of the Upper South. The worn-out hillsides and diminishing population of Virginia, the abandoned farms of Georgia and the Carolinas were proof that the land needed to be taken away from an exclusively one-crop system.[2]

If such a practice continued to prevail in Alabama, Edmund Ruffin was convinced after a trip through the Canebrake region of the State, exhaustive methods of cultivation would lead to the same conditions as were found in the other states. Nor was Ruffin certain that ignorance of fact or want of experience caused such practices to be pursued in Alabama, for scores of Virginians and South Carolinians employed here the same mode of tillage which had exiled them from their native states. The cause appeared to be an indifference to the future of the section as well as an unsettled state of feeling, and until some

---

[1] *Greene County Gazette*, Sept. 20, 1830.

[2] *Ibid.*

means were taken to "fix the population" there could be no hope for a change.[3]

By 1840 the gradual reduction of land prices and the decrease in yield of cotton per acre were indicative of the fact that soil exhaustion was being felt in many parts of Alabama. In Franklin County, where the best farming lands had previously sold for from thirty to sixty dollars per acre, they had now been reduced to an average of from ten to fifteen dollars. In the same county the yield of seed cotton per acre had dropped from 1,000 to 500 pounds.[4] Even in the Black Belt, newspapers, by means of editorials, constantly warned planters of the inevitable results unless they abandoned the one-crop system, for the effects of exhaustion were beginning to be felt by the mid-forties.[5]

Though it is probable that tobacco started the process of soil exhaustion by a succession of crops taken from the same field without rest or intermission in colonial Virginia, agricultural exploitation had reached such proportions in 1815 as to be felt along the whole Atlantic seaboard.[6] Writers urged planters to change their methods of cultivation, to adopt a system of rotation of crops, to raise stock in order to have manure for the worn-out fields, and particularly to plant less cotton and pay more attention to producing their own food.[7] However, the difficulty of using slave labor in a diversified system of farming, and the speculative element which

---

[3] Edmund Ruffin, *Notes on the Cane-Brake Country*, p. 22; *DeBow's Review*, Jan., 1855, p. 26; *The Farmers' Register*, I, 349; *American Cotton Planter*, II, 200.

[4] *Pickensville Register*, June 25, 1842.

[5] *State Journal and Flag*, June 26, 1846.

[6] *The Southern Cultivator*, VIII, 65; *American Cotton Planter*, New Series, IV, 304; Avery O. Craven, "Soil Exhaustion as a Factor in the Agricultural History of Virginia and Maryland", *University of Illinois Studies in Political Science*, XIII, 11.

[7] M. B. Hammond, *The Cotton Industry*, p. 48.

played so large a part in cotton culture, deterred planters from making the desired changes. Drainage and various systems of subsoiling were measures often recommended for deferring, if not preventing, soil exhaustion, but planters made little use of them. The science of deep plowing was little employed, and when shallow cultivation ceased to be profitable the planter usually moved to new lands.[8]

C. C. Langdon, in an address before the Mobile Agricultural and Horticultural Society in 1854, alluded to the fact that planters manifested a strange indifference to anything relating to the preservation and improvement of the soil, the extension and diversification of products and the cultivation of fruits. Year after year they planted their lands in cotton and corn without the application of any fertilizer until the soil finally became worn out and exhausted; then they "pulled up stakes" and, with wife and children, emigrated to some new country where they repeated the same process. Langdon felt that the influence of such a system on the political and social well-being of the State was disastrous, for the exhausted lands would yield support to no one, and the loss in population weakened both the manpower and the wealth of Alabama.[9]

According to the ideas of Professor Craven, there are two distinct problems in the depletion of any soil. One has to do with factors which work immediately upon the soil to lower its yielding capacity, and the other deals with those forces which come with the agricultural system which permits its destruction.[10] One is a problem for the agricultural soils specialist, and the other is work possibly for the sociologist. Since a study of the plantation system involves more precisely the effects of agri-

---

[8] *Ibid.*, p. 82; *The Farmers' Register*, II, 327.

[9] *American Cotton Planter*, II, 258.

[10] Craven, *University of Illinois Studies in Political Science*, XIII, 12.

cultural practices than of chemical reactions, it is with the latter phase that we shall now deal.

The average planter scarcely considered his land anything more than a current expense. He used neither fertilizer nor a system of crop rotation, for his one great aim was to produce cotton, planted year after year in the same fields until the ground refused to bring forth a remunerative crop. If some prospective purchaser looking for land found a field without a stump, he took for granted that it had been worn out.[11] To the man engaged in cotton culture, land was a perishable property, to be worn out and not improved. If it was rich level land, which would last a number of years, the purchaser paid a fair price for it, but if it was rolling, as was most of the land of Alabama, he considered how much the trace was washed or worn and how long the fresh land would last, and in view of these facts he determined the value of the property.[12]

Inasmuch as the most exhausting system of cultivation was especially prevalent in the slaveholding states, the enemies of the institution of slavery contended that slave labor and exhaustive tillage were necessarily connected as cause and effect. Ruffin admitted that slave labor facilitated exhaustion only in the sense that it encouraged one staple production.[13] However, if the energies of slave labor were directed towards improving rather than destroying the productivity of the soil, he believed that results as far-reaching could perhaps have been achieved with slaves as with free labor. Another important factor which led to rapid exhaustion in the cotton states was the fact that the great majority of the crops in the lower South required an unusual amount of tillage. The frequent turning and loosening of the earth by the plow and the hoe year after year without inter-

---

[11] Frederic L. Paxson, *History of the American Frontier*, p. 204.

[12] Hammond, *The Cotton Industry*, p. 83.

[13] *American Cotton Planter*, II, 133.

mission exposed the soil to destructive washing, which was even more accentuated by prolonged periods of heavy rainfall.[14]

C. C. Clay, Jr., in addressing the Chunnennuggee Horticultural Society in 1855, pointed to the effects of soil exhaustion in North Alabama as follows:

> I can show you with sorrow in the older portions of Alabama, and in my native County of Madison, the sad memorials of the artless and exhausting culture of cotton. Our small planters, after taking the cream off their lands, unable to restore them by rest, manures or otherwise, are going further west and south in search of other lands which they may and will despoil and impoverish in like manner. Our wealthy planters, with greater means and no more skill, are buying out their poor neighbors, extending their plantations and adding to their slave force.[15]

Clay further stated that in traversing Madison County one discovered numerous farmhouses once the abode of nonslaveholding whites, but now deserted or turned into slave quarters by the planters who had purchased the land. On all sides lay abandoned fields grown up in broom sedge and cut here and there by gaping gullies. "Indeed, a country in its infancy, where fifty years ago, scarcely a forest tree had been felled by the ax of the pioneer, is already exhibiting painful signs of decay apparent in Virginia and in the Carolinas; the freshness of its agricultural glory is gone, the vigor of its youth is extinct, and the spirit of desolation seems brooding over it."[16]

Generally speaking, soil conservation in Alabama was a luxury far beyond the planter's means. Nevertheless, two planters, Robert Withers and James Thornton of Greene County, were close adherers to the Ruffin school.[17]

---

[14] *Ibid.*

[15] *DeBow's Review*, XIX, 725.

[16] *Ibid.*, XIX, 728.

[17] *The Farmers' Register*, II, 327, 498.

Thornton chastised his overseer for allowing negroes to burn cornstalks in the field, stating that he had moved away from Virginia to get away from impoverished soils and that he wished to prevent soil exhaustion as much as possible in Alabama. The large yields of the "Thornhill" plantation were evidence in themselves that the soil was not ruthlessly depleted.[18] Robert Withers endeavored to prevent soil erosion by a rotation of crops and an adequate system of drainage ditches. This was particularly necessary where the surface was slightly rolling and topsoil could be more rapidly washed away.[19]

Though the agricultural editors repeatedly advocated a rotation of crops, Alabama planters continued to strain their fields almost to the point of exhaustion by one crop after another of cotton and corn.[20] Even though the system began in Tidewater Virginia, the cotton states failed to heed the disastrous results, for it seemed that every agricultural community had to run the same race of exhaustive culture and land impoverishment before being convinced.[21] Many plans of crop rotation were suggested, though all contained the same basic principle in that they allowed part of the plantation to be put into pasture crops each year. One scheme was to divide the total cultivated acreage into four equal parts, planting one in cotton, a second in corn, the third part in grain crops such as wheat or oats, while the fourth was to be used to pasture cattle. For the second year's rotation cotton was to be planted in the portion which was used for pasture and the previous year's cotton fields were to be allowed to remain idle. By this plan cotton would be planted in the same ground only once every four years.[22]

---

18 *Cotton Book of the Thornhill Plantation,* 1850-1858.

19 *The Farmers' Register,* I, 217; W. N. Brewer, "Some Effects of the Plantation System Upon the Ante-Bellum South", *Georgia Historical Quarterly,* XI, 260.

20 *American Cotton Planter,* I, 149.

21 *Ibid.,* II, 130; *The Southern Cultivator,* I, 87-88.

22 *Ibid.,* VII, 162.

Closely associated with the plea for crop rotation came that for diversification. A Black Belt newspaper, in an editorial, stated:

> The practice of farmers of concentrating their attention on a few staple articles such as wheat, cotton, or tobacco is injurious and may be considered one of the causes of the present depression of agricultural interests. Where dependence is placed on a single or even two or three kinds of crops, there is much more probability of loss by a fluctuation of prices than there would be if a greater variety were cultivated; for it seldom occurs that prices of all products of agriculture are depressed by the same cause at the same time. It is without doubt the farmer's best policy to cultivate every article to which his soil and climate are adapted . . .[23]

To make the plantation more self-sustaining by planting less cotton and raising more foodstuffs appears to have been the goal at which all agricultural journalists aimed.[24] Even as early as 1827, newspapers in the Tennessee Valley were suggesting that planters raise their own wheat instead of buying flour at exorbitant prices up the Ohio River.[25] It was generally felt that grain production should at least equal the amount necessary for home use, especially during the periods when the cotton market was low. Cotton production should be curtailed and in its place corn, potatoes, wheat, oats, rye, mules, hogs and cattle substituted. Some even insisted that for a well-planned economy planters should manufacture the clothing necessary for their hands, as well as rope and bagging.[26]

When the cotton market reached a depressed state in the early forties, numerous appeals went forth to reduce the acreage and employ the hands in some other form

---

[23] *Greene County Gazette*, May 31, 1830.

[24] *American Cotton Planter*, New Series, II, 39; *The Southern Cultivator*, II, 123; *DeBow's Review*, I, 434-436.

[25] *Southern Advocate*, May 25, July 27, 1827.

[26] *Huntsville Democrat*, Aug. 13, 1834.

of farming in the hope of arresting the evils of overproduction. It was suicidal to attempt to raise the price of cotton by increasing the number of bales produced. Planters were advised that then was the time to clear new land, repair fences, improve their drainage ditches and concentrate on a more varied agricultural program which embodied laying out orchards and raising more food.[27]

In planning the work on the plantation the planter usually assigned to each field hand from five to eleven acres of cotton, which, if the season was favorable, produced around 2,300 pounds of ginned cotton. The average yield of corn in Alabama was approximately twenty bushels to the acre, but it was actually less work to cultivate twenty acres of corn than eight acres of cotton and six of corn. Consequently, corn production offered a much greater yield per laborer than cotton.[28] However, when the newspapers reported fine crops of cotton in every section, they frequently reported the scarcity and high prices of provisions. Their object seems to have been to persuade the planters of Alabama not to raise large crops of cotton, which inevitably reduced the market price, while at the same time they were paying a dollar a bushel for corn and twelve cents a pound for bacon on the open market.[29] However, the planter invariably sought means of converting more acres into cotton, without stopping to think that his profits were being sliced away. In 1852 a cotton factor in Mobile expressed the opinion that the proceeds of the previous year's crop had already been spent in the purchase of plantation supplies, and that at the rate at which foodstuffs were rising the producer could expect nothing to be left after his current bills were paid.[30]

---

[27] *Southern Advocate,* Dec. 20, 1844.

[28] *State Journal and Flag,* March 20, 1846.

[29] *Sumter County Whig,* Feb. 10, 1852.

[30] *Spirit of the South,* May 25, 1852.

One overseer confessed in 1858 that he had given up the idea of being exclusively a cotton planter and was concentrating on being a farmer in the true sense of the word. In the hope of increasing his profits from a smaller acreage of cotton, he stated, "I have now turned my attention to corn, hogs, peas, potatoes and garden vegetables."[31] In the upper Coosa Valley, Benjamin Yancey, in desperation, turned to a more diversified agricultural program. In 1858 his *Plantation Account Book* indicated that besides 138 bales of cotton there were 3,000 bushels of corn and smaller yields of wheat, rye and barley.[32] Two years later Yancey sold a small portion of his corn crop for $307.05 to people in the neighborhood.[33]

Though to Patrick Henry goes the credit for having remarked soon after the Revolution that "since the achievement of our independence, he is the greatest patriot who stops the most gullies,"[34] Alabama planters appear to have made little effort to improve their lands. Though the use of grasses as soil binders, land retirement, horizontal plowing and hillside ditching were by no means unknown, they were apparently put into practice very little.[35] One great trouble lay in the fact that more land was cultivated than could actually be properly cared for, and surplus cash not needed for the payment of debts was invested in new land instead of being used for improvement of the old. The desire of the planter for the adjoining land of his neighbor was an urge which was seldom, if ever, resisted.[36] The immense

---

[31] *American Cotton Planter*, New Series, II, 113.

[32] Wheat, 210 bushels, rye, 47½ bushels, barley, 21½ bushels.

[33] *B. C. Yancey Coosa River Plantation Account Book*, Jan., 1860.

[34] A. R. Hall, *Early Erosion Control Practices in Virginia*, U. S. Dept. of Agriculture, Misc. Publication No. 256, p. 2.

[35] See *ibid.*, pp. 10, 12, 15, 20 for methods of improving land in Virginia.

[36] *Pickensville Register*, March 12, 1842.

yield capable of being produced on a small acreage of thoroughly tilled land could scarcely be imagined in Alabama, for land in itself was a commodity to be bought and sold as the occasion demanded.[37]

When the new lands of the Southwest had come under full production, depressing the price of cotton to the point at which it could not be profitably grown, restoring the productivity of the soil would probably not have solved the problem. Planters, no doubt, would have continued to increase their production of cotton. There was little hope of acquiring new lands to the south of the United States, and beyond Texas and Missouri the way remained closed. Professor Ramsdell believes that slavery had about reached its zenith by 1860 and must shortly have begun to decline, for the economic forces which carried it across the Mississippi were about to lose their maximum effectiveness. The system was dependent upon an abundant supply of rich land, which is evidence in itself that the slave owner was necessarily an expansionist.[38]

Various estimates were given as to the cost per pound of producing cotton. In 1842 a convention of planters in Mobile agreed unanimously that cotton could not be profitably grown when the price was less than eight cents a pound.[39] On an Alabama plantation, Solon Robinson, a contemporary political economist, stated that when cotton sold for six cents net it would pay expenses, including interest on capital.[40] In 1853, the *American Cotton Planter* estimated the cost per acre of raising cotton as follows:

---

37 *Ibid.*; *American Cotton Planter*, New Series, II, 362.

38 See Charles W. Ramsdell, *Mississippi Valley Historical Review*, XVI, 171.

39 Lewis C. Gray, *History of Agriculture in the Southern United States to 1860*, II, 709.

40 Herbert A. Kellar, ed., *Solon Robinson Pioneer and Agriculturist*, II, 244-245.

Interest on one acre of land worth $25 @ 8%	$ 2.00
Breaking up ground with double team	2.00
Planting seed	1.00
Barring off and scraping	1.50
Ditching and breaking out middles	2.00
Hoeing and scraping	1.00
Picking 1,400 pounds of seed cotton @ 75c	10.50
Ginning and packing	3.00
Hauling to market	1.50
Total	24.50
One bale of cotton (400 lbs.) @ 8 cents	32.00
Net profit	7.50

In the above estimate 400 pounds per acre is used as the average yield, but this is considerably more than most Alabama planters actually produced. Granting that every acre produced 1,100 pounds of seed cotton and that 1,400 pounds were necessary to make a 400 pound bale, five-sevenths of a bale would cost approximately $21.50 to produce, and at eight cents a pound would bring only $22.88. At this figure only $1.38 would be left as net profit.[41]

In estimating the cost of producing cotton, DeBow assumed that the average value of a full field hand was $900 and that a third of the entire force on the plantation consisted of young and old negroes who were incapable of full time work. Thus, in calculating the interest on the value of each hand it was necessary to add 33⅓% to cover the value of the inefficient negroes who formed a necessary part of the cost of cotton planting.[42] If thirty-five acres of land, both cultivated and uncleared, at an average price of $5.30 an acre were allotted each hand, the total interest on land per working hand would amount to $15.00.[43] To this DeBow

[41] *American Cotton Planter,* I, 249.

[42] *DeBow's Review,* April, 1855, p. 469.

[43] In 1850 the average value of farms per acre in Alabama was $5.30.

added one mule for two hands, which with ordinary plantation treatment could be worked for five years and thus must be charged against the expenses of the crop. In addition, the expenses of the overseer, cost of negro clothing, salt, iron and blacksmith work, bagging and rope, annual expense of tools and wagons must be included, making a total of $158.75.[44] If 2,000 pounds of ginned cotton were credited each hand, cotton would cost approximately eight cents a pound.

In the foregoing estimate of the cost of cotton production in Alabama, no charge was made for corn or bacon, which apparently were to be raised at home. If the average production of ginned cotton to the hand was 2,000 pounds, and by prudent economy the planter produced all his own supplies, it actually cost him 7.8 cents for every pound of cotton raised.[45]

Profits in cotton planting in most cases have been largely overestimated. At eight cents per pound, which may perhaps be taken as the average price of cotton, it was a fair business and nothing more. There were individual cases where planters made a large profit at that price, but for the great majority the planting profession meant only a living.[46]

Though a graduated tax ranging from a minimum of twenty-five cents on negroes under five years of age to $1.10 on those between fifteen and thirty, then gradually being reduced to twenty cents a head on those fifty and over, was assessed on slaves in Alabama, there was surprisingly little revenue brought in from this form of taxation.[47] In 1857, William P. Gould paid only $35.65 on fifty-one negroes, while John Erwin with 176 paid

---

[44] *DeBow's Review*, April, 1855, pp. 469-470.

[45] *Ibid.*, p. 470.

[46] *Ibid.*, p. 468.

[47] See Tax Lists of Greene, Perry and Shelby Counties, 1850.

only $231.00.[48] Five years earlier, Gould's total taxes including those on land, slaves, equipment and household furnishings amounted to only $115.14 for the year, and of this sum the State received $95.95 and the county $19.19.[49] In the same year James Tait, who owned five plantations and almost 300 slaves, paid only $421 for State and county taxes.[50]

County records indicate that capital invested in land, plantation equipment and personal property of planters at times reached considerable sums. William Brandon of Madison County in 1848 left 149 slaves valued at $55,700, 1,400 acres of land at $21,000, plantation equipment at $3,000 and personal property at $12,000.[51] In 1853, Joseph Pickens left an estate amounting to over $100,000, but the records indicate that the proceeds from his cotton in that year amounted to only $4,111.72, while the expenses of running his plantation totaled $7,901.92.[52] No doubt, on some of the large estates there was a tendency to overcapitalize, which was natural when both labor and capital were combined as in the case of the slave system, and thus reach the point where the law of diminishing returns set in.

The main object in life for the planter was apparently to raise cotton. There was always the need for an increased acreage and an enlarged slave force to work it. Capital accumulations of profitable years were turned back into the business in the form of more land and more slaves. As a consequence, the planter bought up the farms of his small neighbors and scoured the country for slaves if his own did not breed fast enough.

---

48 *Tax List, Greene County*, 1857. In this year the large slaveholding County of Greene collected only $14,208.30 in taxes on slaves.

49 *Ibid.*, 1852.

50 *James A. Tait Plantation Memorandum Book.*

51 *Estate Files, Madison County*, 1848.

52 *Israel Pickens Papers.*

When the price of cotton rose there was always the temptation to overexpand, and when the market was dull the urge to take advantage of low prices in land and slaves was ever present. No matter what the trend of the market might be, the planting interests of the State were always ready to trade, buy or sell.[53]

Lyell observed that the business of planting in Alabama was beset by the gravest uncertainties.[54] Adverse weather conditions and sudden fluctuations in prices brought financial ruin to many planters. Late spring freshets in the rivers, such as the one which swept through the Tennessee Valley in 1824,[55] tended further to increase the element of chance in planting. In such a setting the planter struggled to overcome the obstacles which were bound up with his pattern of life. An extract from the *Wetumpka Argus* in 1845 sets forth rather clearly his philosophy:

> He wouldn't sell a chicken, nor a dozen eggs, nor a bushel of peaches, nor a colt for any consideration. He is above that! He raises cotton—he does. He rides in a six hundred dollar carriage, for which he is in debt. His daughters strum a piano that never will be paid for. He buys corn which he could raise for ten cents a bushel and pays sixty cents for it, after 2½% is paid to the commission merchant. He could raise his own tobacco yet he pays $3 a pound for Richmond scented. He could raise his own hogs yet he patronizes Cincinnati. The consequences are disastrous. Being the possessor of one staple, he fluctuates with the market of that article. He takes the 'Price Current'—he pays postage, he gobbles down the English news like a cormorant. If he sells today, he'll lose—therefore he'll wait for better advices. He is mixed up in cotton and is a gambler therein. Meantime he wants money—drafts on his factor. He wants cotton goods and clothes for his plantation which he could make at home. He orders them and feels 'large'. The

[53] *Alabama Beacon*, Dec. 3, 1849.

[54] Charles Lyell, *A Second Visit to the United States*, II, 64.

[55] *Huntsville Democrat*, Nov. 2, 1824.

> year closes and he is up to his eye brows in debt. This is a result of his non-calculating, not even guessing, the difference between farmers and planters is that one supports a family and the other supports pride until pride gets a fall.[56]

Nearly everything connected with the plantation was bought on credit. No matter what the price of cotton might be, the planter had to have his supplies. As a consequence, mortgages increased as debts mounted against him, and in many cases, before he died, he was in actuality working for his factor as a result of the accumulated obligations. C. C. Clay, Sr., wrote William F. Withers of Greensboro in 1843, "My debts and liabilities have accumulated until I have no alternative but to submit to legal coercion or sell property at a sacrifice. I prefer the latter, and have been for some months past advertising my old place for sale, without any prospects of success."[57] Clay stated further that he would offer his new place for sale but that he had no idea he could get within half its value in cash. As a consequence, he was undertaking to sell some of his negroes and reduce his living expenses by every practical means in the hopes of paying part of his debts.[58]

William P. Gould was always encumbered with back debts which he had great difficulty in paying. In 1848, an account with Hoadley and Wilson had been unpaid for five years, one with Randall W. Stuart, commission merchant, for $1,690 had been in arrears for four years, and he was three years behind with the account of Dr. Gaston Means.[59] Again in 1850 and 1854 Gould was behind in the payment of his accounts. In the latter year he owed his factor $7,805.16 for plantation supplies,

---

56 *The Southern Cultivator,* III, 58, quoting the *Wetumpka Argus.*

57 *Clay MSS.*, Sept. 16, 1843.

58 Same to same, *ibid.*

59 *William P. Gould Papers,* 1848.

cash drafts and miscellaneous bills, while the proceeds from his cotton crop amounted to only $6,258. The next year Gould noted in his diary:

> I received a letter from Mr. Murray Smith asking me to return the advance which he made me last March, but it is impossible for me to do so. I had counted on obtaining a loan upon my claim on the Hayes estate, but I have been disappointed. The sleepless nights and anxious days that have passed as a consequence, my family well know.[60]

By 1860, James Crawford, commission merchant of Mobile, was pressing Gould for the payment of $2,305.66, which was the accumulation of five years.[61] Local creditors were also urging payment of balances due them, but the expenses of running the "Hill of Howth" plantation continued to increase while the proceeds from cotton tended to remain the same.[62]

As early as 1825, a North Alabama newspaper suggested that perhaps slavery was not as profitable as many people believed. One of the bases for this observation was the exhausted condition of the soil along the Atlantic seaboard, and the great exodus of emigrants from that region to the virgin lands of the Southwest. No doubt, the editor continued, the same conditions which brought the downfall of the slave regime in Virginia would at some future date become apparent in the Tennessee Valley, for it was too great a temptation for the planter not to consider false profits as real profits.[63]

Such a statement at the beginning of the plantation era in Alabama was destined to have far-reaching significance. Several years later, a Black Belt planter, Henry Watson, in writing to a friend, presented an interesting

---

60 *Ibid.*, Diary, Feb. 14, 1855.

61 *William P. Gould Papers*, June, 1860.

62 *Ibid.*

63 *Southern Advocate*, Dec. 30, 1825.

idea as to the outcome of the slave system. Having been asked whether or not he thought slavery would be everlasting he replied, "You ask me if I expect slavery to be perpetuated. I answer that I do not. The slave population increases so fast that this is impossible and I wish some way might be devised which would free us from them . . ."[64] Watson, no doubt, had in mind that when the black population reached the point where the law of diminishing returns set in, negro slavery would cease to be profitable, and then planters would find some means of ridding themselves of their ever-increasing burden.

Highly uncertain returns from cotton production demanded the greatest flexibility, yet the slave system was in nearly every sense rigid. Whenever overproduction depressed the cotton market, the only way of raising the price was to curtail the crop. This, however, was difficult to accomplish, for the planter who owned slaves skilled in the production of cotton found it almost impossible to change to any other crop. Thus to diminish the output required nothing short of heroic efforts, for in the popular estimation the chance for gain overbalanced the probability of loss. Throughout the whole cotton belt, inflation and depression followed each other in rapid succession, but as a usual thing the planter suffered more in the latter period than he gained in the former. Profits were absorbed in the purchasing of additional slaves and in times of stress the planter found it impossible to realize on his investment because of the cost of supplies and the high interest rates.[65]

Lyell observed that more and more planters in Alabama were being convinced that slave labor was more expensive than free labor. "Half the population . . . is employed to see that the other half do their work,

---

64 Henry Watson, Jr., to Julius Reed, July 23, 1834, *Watson MSS.*

65 See U. B. Phillips in the *Political Science Quarterly*, XX, 274.

and they who do work accomplish half what they might do under a better system."[66] Only in rare instances could planters raise enough capital to establish their own cotton mills because all their savings went to buy negroes, or in depressed times to feed those they already had.[67] James Tait estimated that the cost of maintaining a negro on his plantations amounted to $34.70 a year, which went for food, clothes, medical attention and taxes,[68] to which had to be added interest on the money invested in the slave, time lost from work because of sickness, and other incidental expenses.

In the early days of Alabama, certain industrialists bought slaves and attempted to operate with their own labor, but later they came to abandon the scheme as unprofitable. By 1840, railroads had definitely given up the attempt to use slave labor for track maintenance crews because free labor in Alabama could be hired for from fifty to sixty dollars a year and, considering the type work involved, the loss of time because of sickness and the initial cost of the investment, slave labor was considered more expensive.[69]

Cotton, however, was not the only source of profit, for a planter barely making expenses might sell some of his surplus negroes. John W. Walker of Madison County in 1832 realized only seventy-six dollars net profit on his cotton crop after deducting the current expenses from the gross sales. This, however, was somewhat offset by the sale of a few negroes to planters in the neighborhood.[70] Another planter in Montgomery County found that his small profits from the sale of cotton could be increased by a system of diversified farming. One-half of his 600 acres of tillable land was

---

[66] Lyell, *A Second Visit to the United States*, II, 72.

[67] *Ibid.*

[68] *James A. Tait Plantation Memorandum Book*, 1851.

[69] *DeBow's Review*, March, 1850, p. 271.

[70] *John W. Walker Papers.*

planted in cotton while the remainder was devoted to corn, fodder, oats, peas, potatoes, and the raising of hogs. Food to feed his forty field hands was produced at home while the remainder was disposed of in the Montgomery market. Between the years 1850 and 1854 an average of $2,500 per year was earned, which amounted to approximately sixty dollars per negro, in addition to the amount brought in by the cotton crop.[71]

In most instances, however, the plan of diversified farming was little subscribed to in Alabama, and the one-crop system continued to be perpetuated until it was interrupted by the war. On numerous occasions, agricultural meetings were held for the purpose of curtailing production and thus raising the price of cotton, but little if anything was ever accomplished. Such a meeting, held in Montgomery in 1845, resolved to enforce pledges from planters to reduce their cotton acreages and concentrate their energy on a more diversified agricultural system, to encourage the manufacture of cotton goods in Alabama, and to petition the legislature for a geological survey of the State in order that water power and mineral resources might be developed. Finally, a recommendation was made for the formation of agricultural societies to encourage domestic economy and to direct labor and capital into their most profitable channels.[72]

Though the conventions pointed out definite problems in the agricultural life of the State, there is little evidence to show that the attending members took them very seriously after the meetings adjourned. Perhaps each was convinced in his own mind that his neighbor would return home and reduce the acreage of his cotton crop, thereby increasing the price. Therefore, to those who increased their individual yields would fall the in-

---

71 *American Cotton Planter*, III, 150-151.

72 *Southern Advocate*, Feb. 28, 1845.

creased profits brought about by a rising market. Thus the planter, spurred on by the spectre of multiplied profits, played the part that was expected of him and increased his acres in cotton with all the means at his command. When the market slumped, as it inevitably did, he was usually the first to lay the blame where it least belonged.

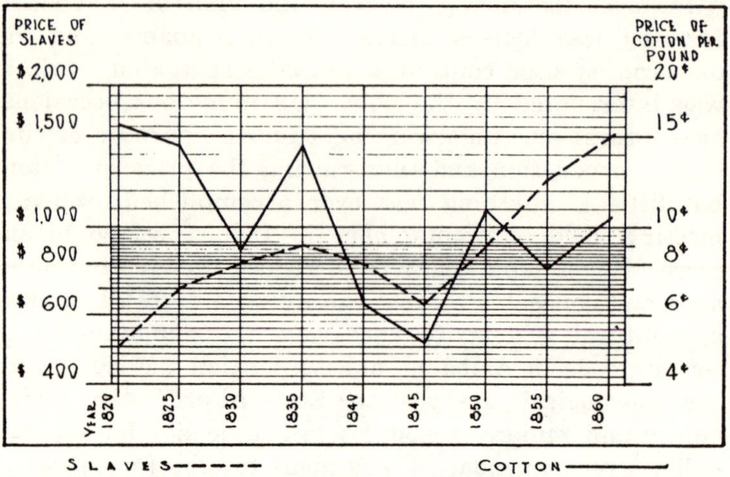

COMPARISON OF THE PRICES OF COTTON AND SLAVES IN ALABAMA

The two periods of greatest profit for the Alabama planter seem to have been from 1817 to 1825 and during the late fifties. In the first period the price of slaves was low and that of cotton high, while in the second the price of slaves had risen far higher than the proportionate value of cotton. Many planters, however, had already acquired the greater part of their slave forces, and perhaps improved methods of plantation management tended to offset somewhat the low price of cotton. The price of slaves rose from $600 in 1820 to $1,600 in 1860, and simultaneously the price of cotton fell during the same period from twenty cents to ten or twelve. The price of both cotton and slaves fluctuated throughout the whole ante-bellum period. For a great part of the time there was a definite correlation between the two, though during the last five years before the Civil War the price of slaves forged far ahead.[73] One might justly conclude that since slave prices were so high, the Alabama planter could receive only a small return on his investment, and that only by efficient management in the cultivation of good lands.

Perhaps the profits in planting have in most instances been exaggerated, but there is no doubt that certain Alabama producers made considerable money in the production of cotton. Profits, however, depended upon the cultivation of rich lands as well as the efficient management of both capital and labor. Even the planters themselves were in agreement that a large part of the food supply had to be raised at home in order to preserve the slender margin of profit which inevitably resulted because of certain fixed charges which had to be met. Nevertheless, cotton continued to receive the major emphasis of all the plantation activities, and from the markets in Mobile and New Orleans came a considerable portion of the food supply. The plantation system also hastened the exhaustion of soil in most of the State's best planting regions, for even at the present time, with the proper use of commercial fertilizer, the yield of cotton per acre in the Black Belt is none too large. From the point of view of economics it is not altogether certain whether or not slavery was a sound investment in Alabama considering the ante-bellum period as a whole, for the forces which would have made its downfall inevitable had long been at work, only to be cut short by emancipation.[74]

---

[73] *Tuskegee Republican*, March 25, 1858.

[74] See Charles S. Sydnor, *Slavery in Mississippi*, Ch. VIII, for good account of the profitableness of slavery.

APPENDIX A

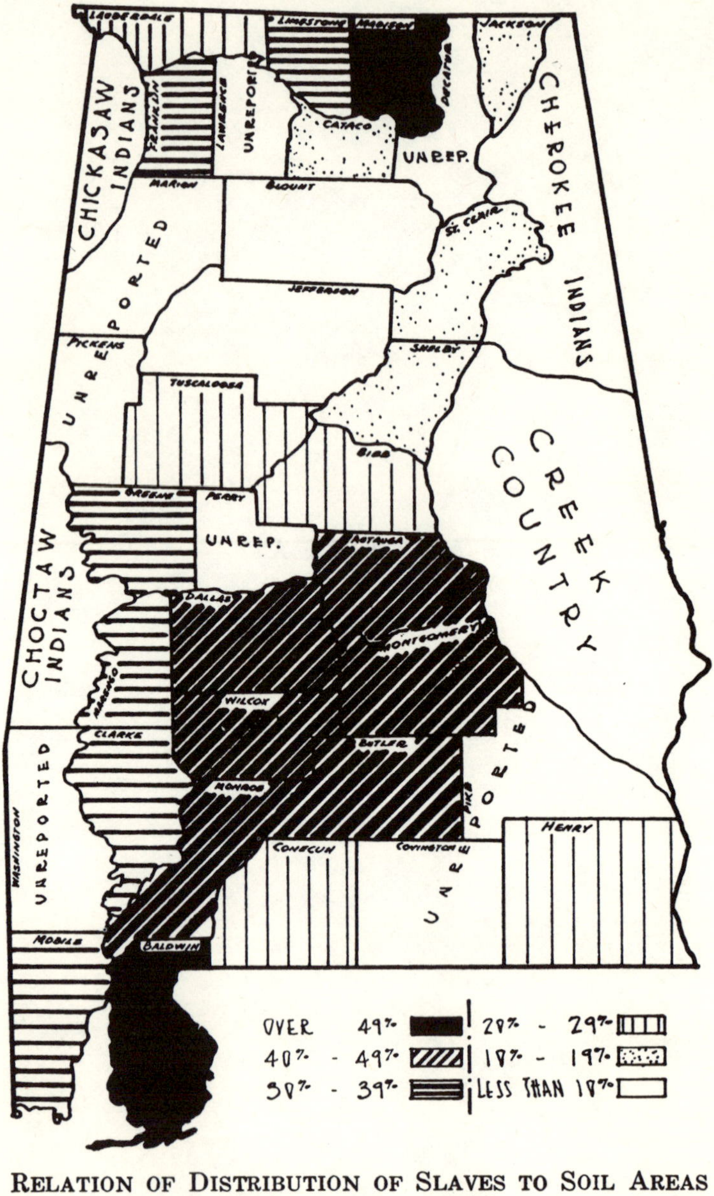

RELATION OF DISTRIBUTION OF SLAVES TO SOIL AREAS IN ALABAMA — 1820

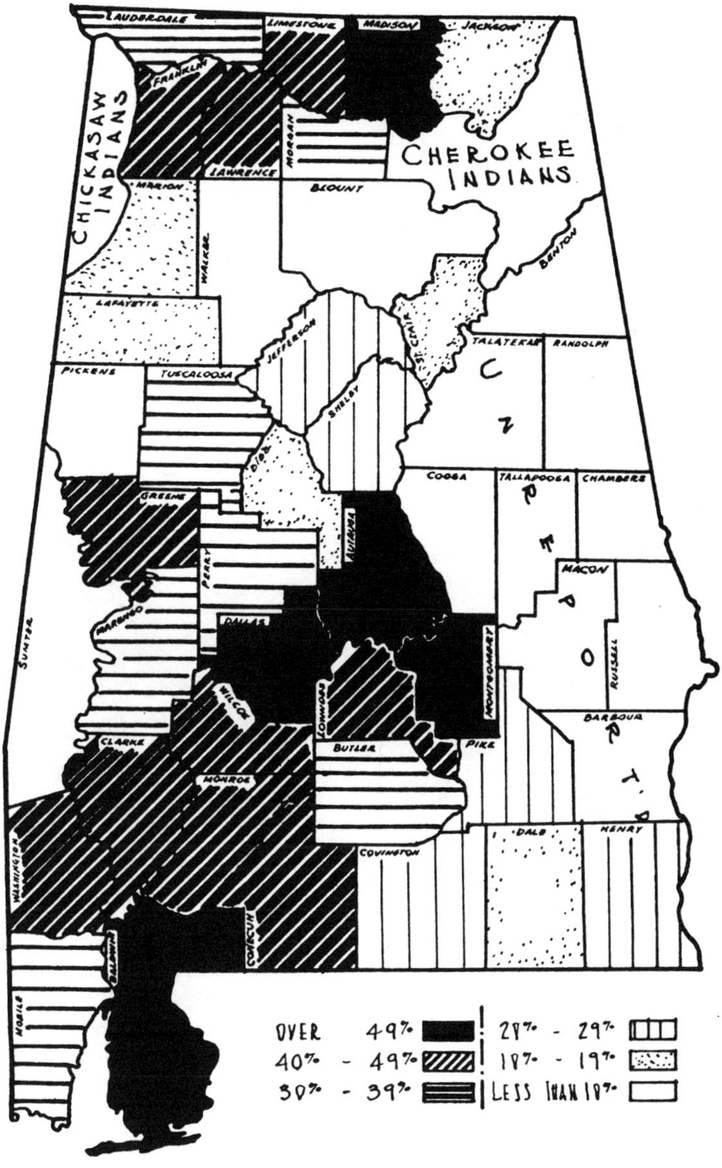

RELATION OF DISTRIBUTION OF SLAVES TO SOIL AREAS IN ALABAMA — 1830

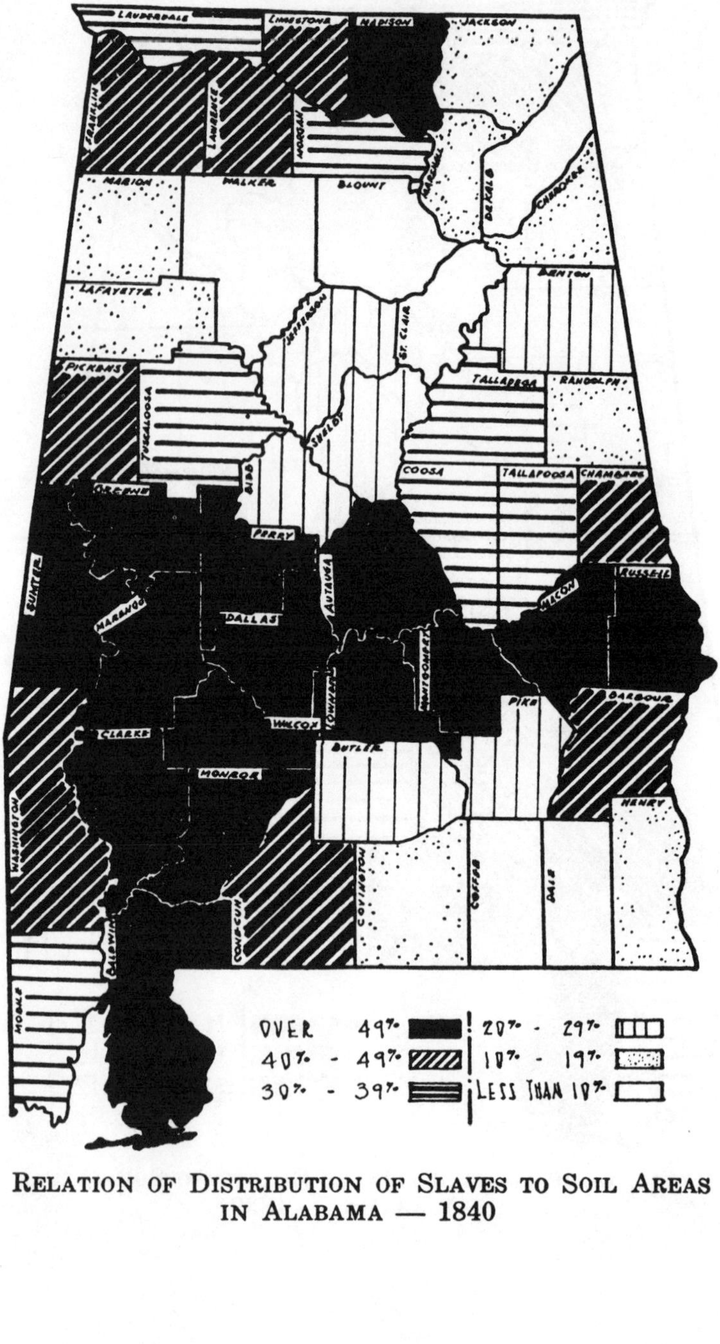

RELATION OF DISTRIBUTION OF SLAVES TO SOIL AREAS IN ALABAMA — 1840

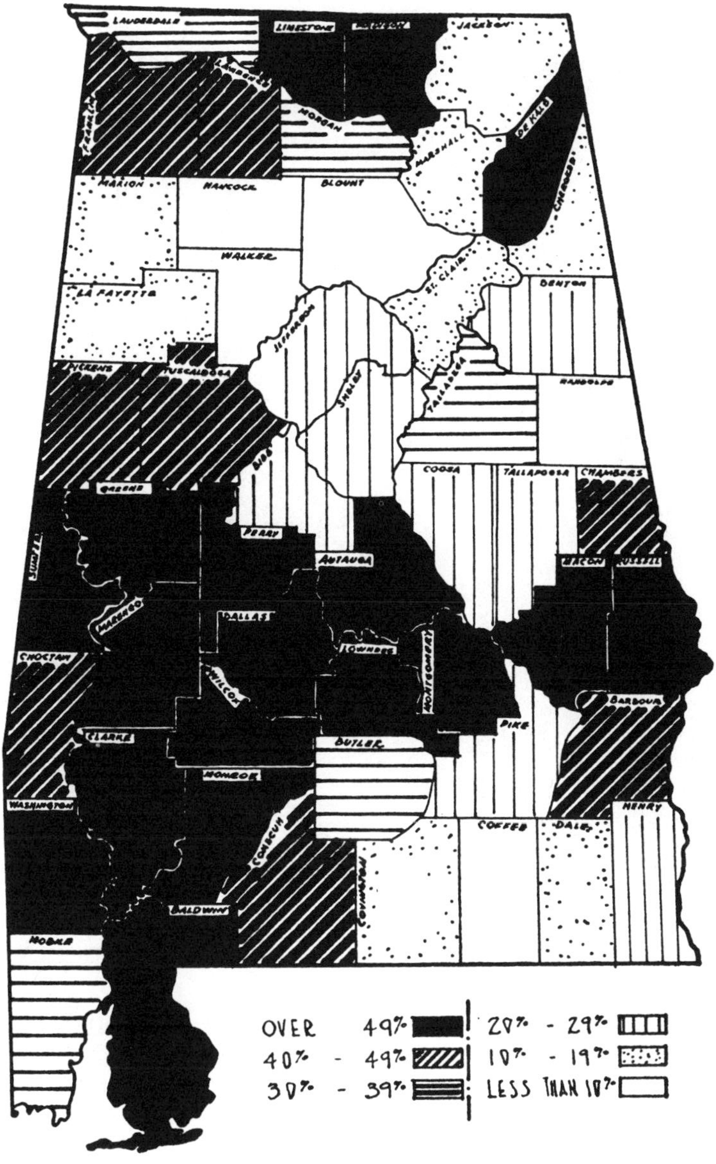

RELATION OF DISTRIBUTION OF SLAVES TO SOIL AREAS IN ALABAMA — 1850

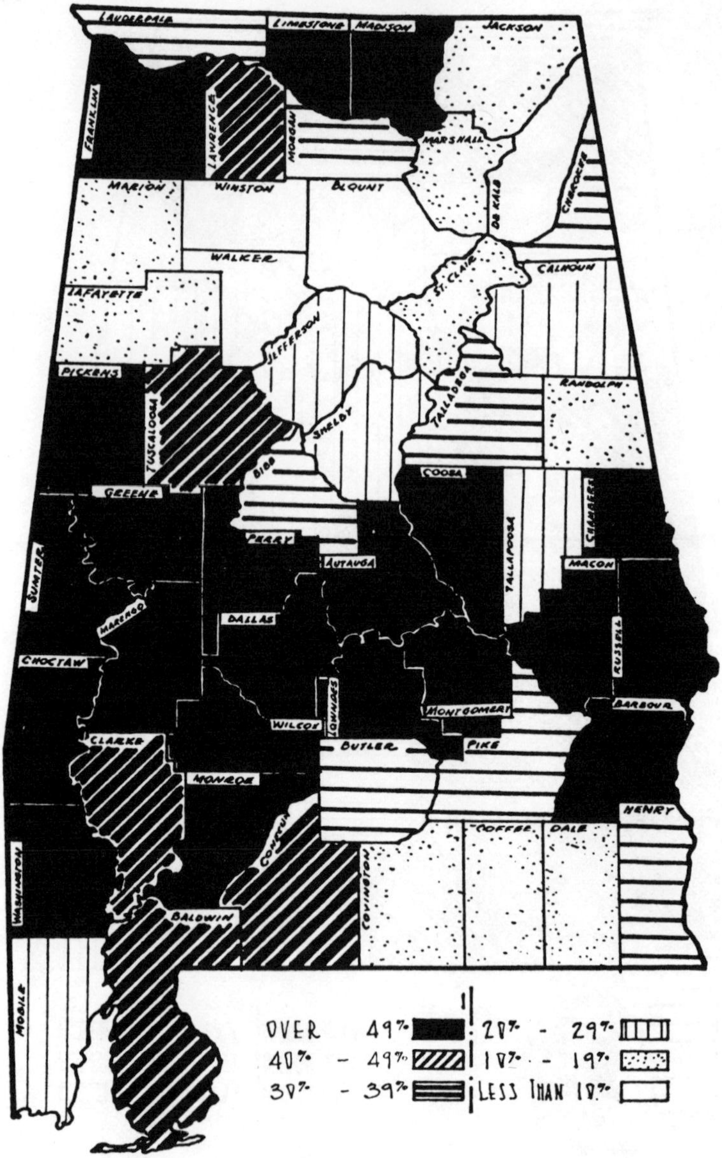

RELATION OF DISTRIBUTION OF SLAVES TO SOIL AREAS IN ALABAMA — 1860

APPENDIX B

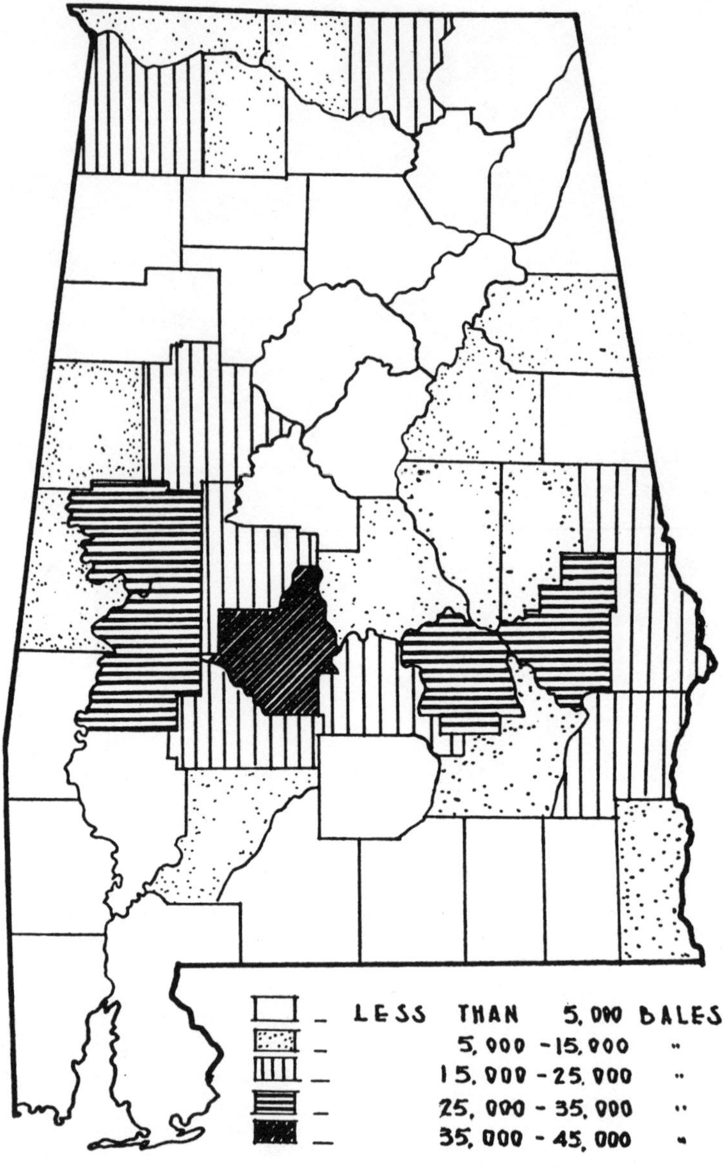

DISTRIBUTION OF COTTON PRODUCTION IN ALABAMA—1850

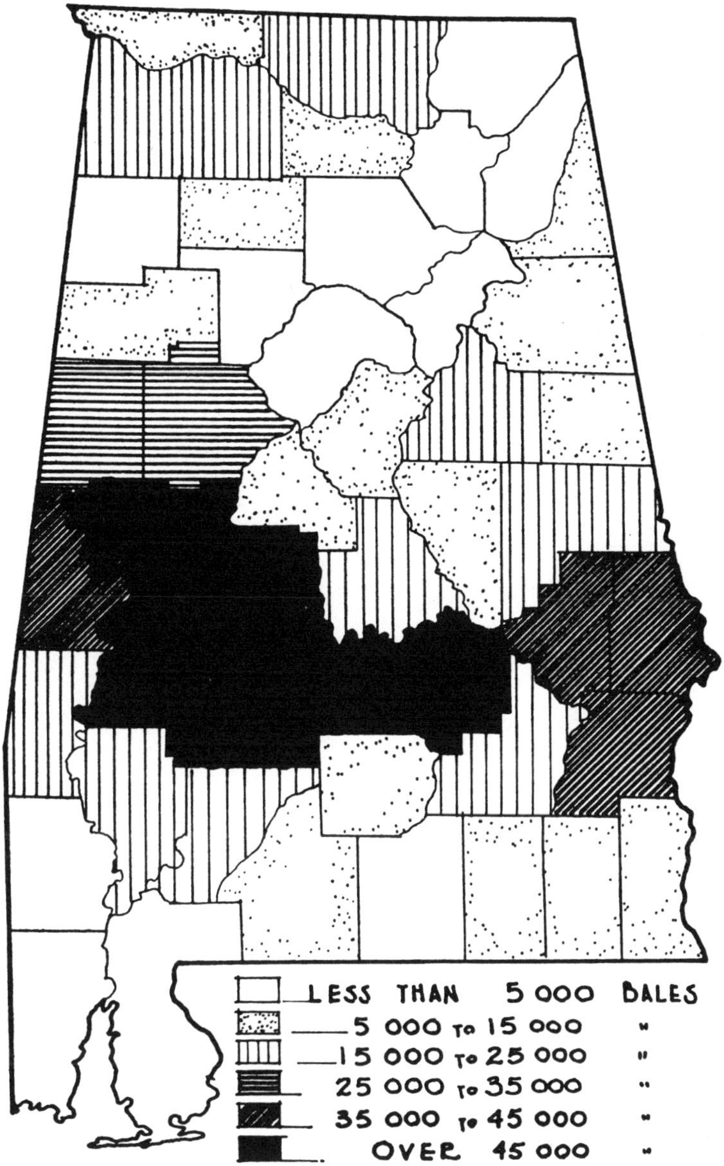

DISTRIBUTION OF COTTON PRODUCTION IN ALABAMA—1860

BIBLIOGRAPHY

Manuscripts

I. Alabama State Department of Archives and History:

Account Book of C. J. W., Wetumpka, Alabama, 1857-1860. This account book is valuable for information concerning the small-town merchant as an intermediate cotton factor.

Assessment of Taxes on Real Estate in the County of Sumter for the Year 1850.

Assessment of Taxes on Real Estate in the County of Tuscaloosa, 1848.

Charles Crommelin Papers, 1845.
This collection contains instructions of Crommelin to his overseers.

Matthew P. Blue Papers.
These papers consist chiefly of information concerning the early history of several Alabama counties, various men having contributed from their personal knowledge.

J. M. Bradford Store Account Books, Wetumpka, Alabama, 1837-1850.
Interesting among Bradford's accounts were numerous purchases of small amounts of cotton.

Ledger of Bradford and Shotwell, Springville, St. Clair County, 1837.
These records are useful for obtaining information as to commodity prices in a small slaveholding county.

Thomas Brown Journal.

This journal contains a few miscellaneous items concerning plantation affairs.

William P. Brown Papers, 1835-1860.

This collection deals principally with early mining operations in Alabama, but sheds considerable light on early transportation problems.

A. A. Causby Account Book, 1839.

This account book contains scattering information regarding prices of staples.

Chronicles of the Canebrake.

These are the personal reminiscences of J. W. DuBose concerning plantation life in the Black Belt.

Manuscript Census Returns for the State of Alabama, 1820 and 1850 for the following counties: Autauga, Baldwin, Barbour, Blount, Chambers, Choctaw, Clarke, Coosa, Dallas, Franklin, Greene, Hancock, Henry, Lawrence, Limestone, Lowndes, Macon, Madison, Marengo, Marion, Marshall, Mobile, Montgomery, Monroe, Morgan, Perry, Shelby, St. Clair, Sumter and Wilcox.

John Coffee Papers, 1815-1850.

These papers are valuable for information on public lands, transportation, and general facts regarding plantation routine in the Tennessee Valley.

Discount Book of the Mobile Branch Bank, 1839-1841.

This discount book is enlightening as to the amount and ownership of notes discounted by the bank.

Final Record of the Circuit Court of Cotaco County, 1830.

Tax List of Greene County, 1849.

Bolling Hall Manuscripts.

This collection reveals valuable information on politics, immigration and early planting activities in the State.

Lauderdale County Chancery Records, Volume C.

Records of the Commissioners Court, Lawrence County, 1818-1824.

Ledger of C. A. Adkinson, Marion, Alabama, 1823-1825.

Ledger of William P. Mollett, Providence, 1854-1869.

Letter Books of the Mobile Branch Bank, 1846-1856.

Lists of Overseers and Roads in Lauderdale County, 1844.

Records of the Madison County Superior Court, 1811-1818.

Colin J. McRae Papers.

This collection contains several letters which give information concerning the cotton commission business.

Minute Book of the Branch Bank of Montgomery, 1838-1843.

Records of the Jefferson County Circuit Court, 1820-1825.

Records of the United States Land Offices in Alabama.

These are almost complete records of the activities of the United States Land Offices from their beginnings to 1860. The collection consists of Tract Books, Receivers Balance Account Books, Letter Books, Sales Books, Cash Entries Books, Registers of Certificates, and others.

Israel Pickens Papers.

This collection of letters and plantation accounts gives helpful information on plantation routine.

Samuel Pickens Papers.

These papers are valuable for information concerning factorage connections of Black Belt planters with Mobile. The collection also contains materials concerning plantation operations in Greene County.

Albert J. Pickett Papers.

These papers contain certain unpublished bits of information dealing with the early history of the State.

Proceedings of the Town Council of Hayneville, Lowndes County, 1836-1842.

This manuscript reveals enlightening information as to town ordinances concerning slaves.

Shelby County Tax List (Personal Property), 1850.

Store Account Book of J. B. Peabody, Huntsville, 1833-1838.

Supreme Court Docket, 1821-1825.

Charles Tait Papers.

These are a valuable collection of plantation journals, diaries and correspondence which is particularly useful for information pertaining to the beginning of the plantation era in Alabama, 1818-1835.

Transactions of the Chunnennuggee Horticultural Society, 1847-1853.

John W. Walker Papers.

These are useful principally for certain statistical information pertaining to North Alabama, 1818-1835.

Cotton Warehouse Receipt Book of William P. Mollett, Providence, 1856-1858.

Diary of Dr. H. V. Wooten, 1813-1837.

This diary contains an excellent detailed account of a frontier doctor.

II. LIBRARY OF CONGRESS:

Diary of Edmund Ruffin, 1856-1865, 25 volumes.

Benjamin C. Yancey Coosa River Plantation Account Book, 1842-1860.

This account book contains important information concerning plantation operations, factorage connections, and the management of negroes in Cherokee County.

III. NEW YORK PUBLIC LIBRARY:

Dubois, Diary of a Journey from New York to Mobile, Alabama, 1840-1843.

Contract for the Sale of Slaves between Rachel English and William L. Crawford of Monroe County, Alabama, March, 1837.

IV. LIBRARY OF DUKE UNIVERSITY:

Clement C. Clay Manuscripts, 1811-1925.
This collection contains several thousand pieces pertaining principally to State and national politics, but there are also miscellaneous bits of information concerning cotton planting and plantation life in the Tennessee Valley.

Jones Fuller Letters and Papers, 1810-1895, Lewisburg, N. C.
Among these papers are several letters with instructions to purchase medicines and giving methods of treating negroes for fevers. In them is also found information concerning transportation of cotton in Alabama.

James S. Gaines Letters, Sullivan County, Tennessee, 1823-1876.
Among these letters there are a few containing information about Tennessee Valley plantations.

James T. Gee Manuscripts, 1837-1864.
Gee's letters to his son at the University of Virginia give information about the hard times brought about by the low price of cotton.

William H. Hatchett Letters, Lunenburg, Virginia, 1828-1849.
These letters are useful principally for facts concerning slave-trading activities in Alabama.

Henry Watson Letters and Papers, Greensboro, Alabama, 1765-1888.
These are letters and papers dealing with plantation management, factorage connections, and adversities in planting and contribute useful information for the Black Belt.

V. LOCAL ARCHIVES:

Dallas County:
Record of Deeds and Mortgages.
Will Record Books.

Accounts, Administrators, Guardianships, Inventories and Appraisements.

Coosa County:

Will Books and Deed Books.

Greene County:

Deed Records.

Orphans Court Records.

Will Books.

Probate Records.

Real and Personal Property Tax Books.

Limestone County:

Deed Records and Will Books.

Madison County:

Files of Estates.

Deed Records.

Will Records.

Probate Records.

Minutes of the Orphans Court.

Perry County:

Will Books.

Assessment Books for Real and Personal Property.

Talladega County:

Probate Minute Books.

Register of Wills and Inventories.

Register of Wills (Old Series).

Deed Records.

Contract between W. P. Bocock and Lewis O. Collins, Overseer, Marengo County, 1860, Birmingham Public Library. (Photostat)

Reports and Manifests, 1806-1811, Customs House, Mobile, Alabama.

Hugh Rolston Manuscripts, 1795-1815, Mobile Public Library.

The manuscripts are useful for the information they contain on early commercial activity and slave trade at Mobile.

VI. RECORDS IN PRIVATE POSSESSION:

L. M. Burford Manuscripts.
These are letters and papers of a fragmentary nature regarding planting operations in Wilcox County. A diary covering the eighteen-fifties is perhaps the most important item in the collection. In private possession.

Store Account of the Canton Community Store, Wilcox County, 1829.
This account book gives useful information as to commodity prices of crossroads merchants. In private possession.

John Coffee Papers, 1820-1830.
These papers consist of personal letters in the possession of Coffee's descendants, Huntsville, Alabama.

A. A. Coleman Plantation Memorandum Book.
This memorandum book contains information regarding plantation management in Sumter County. The manuscript is in private possession in Birmingham.

Account Book of Peter Downey, Jefferson County, 1857-1859.
Entries in this account book indicate that small amounts of cotton were being exchanged for staple groceries by the nonslaveholding farmers. In private possession, Birmingham, Alabama.

Extract from the Journal of John Eads and Family.
This manuscript contains information about Georgia emigration to Clarke County in 1818. In possession of the author.

Will of R. H. Gregg, Peachtree, Alabama, 1860.
In private possession.

William P. Gould Papers.
Gould kept in his diaries a detailed record of his planting activities from 1815 to 1862. The col-

lection also consists of a large amount of business and private correspondence through the whole period, covering every phase of the planting profession. This was probably the most important single collection to which the author had access. In private possession, Greene County, Alabama.

Warehouse Ledger of C. Hall and Company, 1859-1861.
This manuscript indicates the monthly average of plantation supplies arriving from Mobile purchased by planters in the Tilden community. In possession of the author.

David E. Kelley Papers.
These papers, relative to transportation in the Tennessee Valley, are in the possession of Mrs. D. E. Kelley, Madison County, Alabama.

John Richardson Mason Papers.
These papers are important for information concerning the marketing of Tennessee Valley cotton in Chattanooga. Now in the possession of Miss Mary Mason, Athens, Alabama.

Frank Mathews Papers.
These are useful for information concerning overseers on several Black Belt plantations. In private possession.

Alexander Pope Manuscripts.
This collection consists of account books, papers and correspondence of a prominent Mobile commission merchant during the thirties and forties. It is particularly valuable for information regarding the business connections of the Mobile factor with British buyers. In private possession.

Josiah Robins Papers.
These papers are important chiefly for the record of a single sale of negroes to Robins by a Maryland slave dealer involving $32,000. The papers are now in private possession.

James A. Tait Papers.

The plantation memorandum books and business correspondence of Tait convey in detail a tremendous amount of information on every phase of plantation operations. In private possession.

Cotton Book of the Thorn Hill Plantation, 1850.

This book was actually a daily account book of James Thornton's overseer, containing various items of plantation routine. In the possession of Mr. Ennis Thornton, Watsonia, Alabama.

M. Harris Walker Papers.

The slave books indicate the amount of work allotted daily to each slave, in addition to containing bits of miscellaneous information concerning the health of negroes on the Walker plantation. In private possession.

Medical Account Book of Dr. Walker of Dallas County, 1842-1860.

This account book is interesting for the facts concerning fees charged planters for attending negroes in south Dallas County. In private possession.

Invoice Book of Bidgood and Walton, Forkland, Alabama.

This is valuable as a price index for the inland merchant for the years 1815-1830. Now in the possession of the Rosemount Estate, Greene County, Alabama.

Store Account Book, Forkland, Alabama, 1839.

This is a continuation of the business records of Bidgood and Walton. Now in the possession of the Rosemount Estate, Greene County, Alabama.

Tax List of the I. G. Walton Lands, 1850.

In possession of the Rosemount Estate, Greene County, Alabama.

William Walton Papers, 1840-1860.

These papers are important for information regarding sales of public lands in West Alabama. In possession of the Rosemount Estate, Greene County, Alabama.

Ledger and Account Books of Joseph W. Wilson, 1831-1845.

Wilson, a cotton factor and merchant of Montgomery, did a considerable forwarding business with small planters. His books give information as to the volume of business, profits and prices during the period. Now in the possession of Miss Lucharlle Wilson, Montgomery, Alabama.

PRINTED SOURCES

I. PUBLIC RECORDS:

Acts of the General Assembly of the State of Alabama, 1819-1860 (The place of publication varies with the location of the capital).

Aikin, John G., *A Digest of the Laws of the State of Alabama*, Mobile, 1836.

American State Papers, Washington, 1848-1861. 8 vols.

Annual Reports of the Commissioners of Patents, 1845-1860, Washington (Various dates).

Clay, C. C., *A Digest of the Laws of the State of Alabama*, Tuskaloosa, 1843.

Collection of the Ordinances now in force in the City of Mobile, Mobile, 1835.

A Digest of the Various Acts of the Legislature incorporating the City of Mobile together with Special and Private Acts, Mobile, 1849.

Hitchcock, Henry, *The Alabama Justice of the Peace Together with the Constitution of the State of Alabama*, Cahawba, 1822.

House Executive Documents, 24th Cong., 1st Sess., 1835-1836, Docs. 114-197, Washington, 1836.

Mobile City Directory, Mobile, 1837.

Reports of the Supreme Court of the State of Alabama:
Minor's Reports, 1820-1826, New York and Atlanta.
Stewart's Reports, 1827-1831, Montgomery.
Stewart's and Porter's Reports, 1831-1834, Tuscaloosa.
Porter's Reports, 1834-1839, Tuscaloosa.
Alabama Reports, 1840-1860, Tuscaloosa and Montgomery.

Sanford, John W. A., *The Code of the City of Montgomery*, Montgomery, 1861.

Semple, Henry C., *The Code of Alabama*, Montgomery, 1852.

Shepherd, J. W., *Digest of Alabama Reports*, Montgomery, 1858.

Statutes of the Mississippi Territory, Natchez, 1816.

Toulmin, Harry, *A Digest of the Laws of the State of Alabama*, Cahawba, 1819.

II. TRAVEL BOOKS:

The Travels of William Bartram, n. p. (Reprint), 1928.

Bossu, M., *Nouveaux Voyages aux Indes Occidentales*, Paris, 1768.

Buckingham, J. S., *The Slave States of America*, London, 1842. 2 vols.

Darby, William, *The Emigrants Guide*, New York, 1818.

Dow, Rev. Lorenzo, *History of the Cosmopolite or Lorenzo's Journal*, Cincinnati, 1855.

Fuller, Hiram, *Belle Brittan on a Tour, at Newport and Here and There*, New York, 1858.

Gosse, Philip Henry, *Letters from Alabama*, London, 1859.

Hall, Captain Basil, *Travels in North America in the Years 1827-1828*, Edinburgh, 1829. 3 vols.

Hodgson, Adam, *Letters from North America*, London, 1824. 2 vols.

Ingraham, J. H., ed., *The Sunny South, or The Southerner at Home,* Philadelphia, 1860.

Kellar, Herbert A., ed., *Solon Robinson Pioneer and Agriculturist,* Indianapolis, 1936. 2 vols.

Lanman, Charles, *Adventures in the Wilds of the United States and British American Provinces,* Philadelphia, 1856. 2 vols.

Lavasseur, A., *Lafayette en Amerique,* Bruselles, 1829. 3 vols.

Lyell, Sir Charles, *A Second Visit to the United States of America,* New York, 1850. 2 vols.

Martineau, Harriet, *Society in America,* New York, 1837. 2 vols.

Olmsted, Frederick Law, *A Journey in the Back Country,* New York, 1863.

Olmsted, Frederick Law, *The Cotton Kingdom,* New York, 1861. 2 vols.

Olmsted, Frederick Law, *A Journey in the Seaboard Slave States,* New York, 1856.

Royal, Mrs. Annie, *Sketches of History, Life and Manners in the United States,* New Haven, 1826.

Royal, Annie, *Letters from Alabama,* Washington, 1830.

Stuart, James, *Three Years in North America,* New York, 1833. 2 vols.

III. PAMPHLETS:

A Condensed Statement of the Indebtedness of the Several Counties in the State of Alabama to the Bank of the State of Alabama, Tuscaloosa, 1844.

Contemporary Sketch of Florence, n. p., n. d.

Mr. Faulkner, of Virginia, on the Land Policy of the United States, Washington, 1857.

Letter from the Secretary of the Treasury Transmitting a Statement of the Sales of Public Lands in the Alabama Territory, Washington, 1818.

Croom, I., *A Memoir, On the Subject of the Cotton-plant, Its History, Influence on Commerce, Politics, and the Welfare of the Human Race, and its Probable Destiny as the Great Product of the Southern United States,* Tuscaloosa, 1852.

Memorial and Proceedings of the River and Harbor Improvement Convention Assembled at Tuscaloosa, November 17, 1885, Cincinnati, 1886.

Proceedings of the Twelfth Annual Meeting of the Stockholders of the Mobile and Ohio Railroad Company Held in Mobile February 20, 1860, Mobile, 1860.

Ravesies, Paul, *Scenes and Settlers of Alabama,* n. p., n. d.

Reports of the President, Directors and Chief Engineer of the Alabama and Tennessee Rivers Rail Road Company, Montgomery, 1851.

Reports of the President and Directors of the Alabama and Tennessee Rivers Railroad, Selma, 1854.

Report of the Commissioners to Examine the Bank of Mobile, Mobile, 1855.

Report of the Chief Engineer to the President and Board of Directors of the South and North Alabama Railroad Company, Montgomery, 1859.

Ruffin, Edmund, *Notes on the Cane-Brake Lands or the Cretaceous Region of Alabama* (Reprint from the *Southern Planter,* 1858), Rare Book Collection, Library of Cong.

Slade, M. D. J., *The Plea and the Opinions of the Judges of the Supreme Court of Alabama Concerning the Bank of the State of Alabama vs. J. M. Bates et al.,* Tuscaloosa, 1841.

Stein, Albert, *Report on the Improvement of the Alabama River,* Mobile, 1855.

Transactions of the Alabama Historical Society at the Annual Meeting in the City of Tuscaloosa July 9 and 10, 1855, Tuscaloosa, 1855.

Weston, George M., *The Poor Whites of the South,* Washington, 1856.

IV. NEWSPAPERS:

Alabama Beacon, 1843-1860 (Greensboro).
Alabama Baptist Advocate, 1849-1850 (Marion).
Alabama Planter, 1848-1849 (Mobile).
Alabama Journal, 1826-1827 (Montgomery).
Alabama Republican, 1820-1822 (Huntsville).
Alabama Sentinel, March, 1826-July, 1826 (Tuscaloosa).
Alabama Sentinel, April, 1836-September, 1836 (Greensboro).
Alabama State Intelligencer, 1829-1830 (Tuscaloosa).
The Blakely Sun and Alabama Advertiser, 1819 (Blakely).
Banner of Reform, 1840 (Mobile).
Clarke County Democrat, 1856-1860 (Grove Hill).
Clarke County Post, 1836-1837 (Suggsville).
Daily State Guard, 1849 (Elmore County).
Daily Messenger, 1856-1857 (Montgomery).
Dallas Gazette, 1854-1857 (Cahaba).
Dorsey's Dispatch, 1856-1857 (Wetumpka).
Eufaula Democrat, 1846-1851.
Flag of the Union, Jan., 1841-Dec., 1841 (Tuscaloosa).
Florence Gazette, 1858-1860.
Franklin Enquirer, March, 1824-June, 1824 (Tuscumbia).
Greene County Gazette, 1830-1832 (Erie).
Huntsville Democrat, 1823-1856.
The Independent, 1854-1858 (Gainesville).
Independent American, 1855-1859 (Troy).
Independent Monitor, 1847-1860 (Tuscaloosa).
Jacksonville Republican, 1840-1851.
Macon Republican, 1849-1856 (Tuskegee).
Mobile Advertiser and Chronicle, 1839-1840.
Mobile Commercial Register, 1821-1834.
Mobile Commercial Register and Patriot, 1832-1840.

The Mobile Daily Register, 1849-1855.
The Mobile Daily Advertiser, 1843-1856.
Mobile Daily Commercial Register and Patriot, 1836-1841.
The Mobile Mercantile Advertiser, 1833-1839.
Mobile Morning Chronicle, 1835-1836.
Mobile Register and Journal, 1841-1849.
The Mobile Weekly Advertiser, 1854-1855.
Montgomery Advertiser, 1854-1860.
Miscellaneous Files of Montgomery Newspapers, 1837-1854.
Moulton Democrat, 1855-1858 (Lawrence County).
North Alabamian, 1836-1848 (Tuscumbia).
Pickensville Register, 1842-1843.
Spirit of the South, 1851-1852 (Eufaula).
South Western Baptist, 1851-1860 (Tuskegee).
Southern Advocate, 1825-1860 (Huntsville).
State Journal and Flag, Jan., 1846-Nov., 1846 (Tuscaloosa).
States Rights Democrat, Sept., 1860-Nov., 1860 (Jacksonville).
Selma Free Press, 1835-1841.
Sumter County Whig, 1851-1854 (Livingston).
The Telegraph and Patriot, May, 1828-July, 1828 (Tuscumbia).
Tuscumbia Telegraph, 1827-1828.
The Tuscumbian, 1824-1827.
Tuskegee Republican, 1856-1859.
Wetumpka Argus and Commercial Advertiser, 1839-1842.
Wetumpka Spectator, 1856-1857.
Universalist Herald, 1857-1860 (Montgomery).

V. CONTEMPORARY PERIODICALS:

The American Cotton Planter, Montgomery, Ala., 1853-1860.
American Quarterly Register, Boston, Mass., 1827-1843.

DeBow, J. D. B., ed., *The Commercial Review of the South and West,* New Orleans, La., 1846-1880.

The Cultivator, Albany, N. Y., 1834-1852.

Hunt's Merchants' Magazine and Commercial Review, New York, N. Y., 1839-1870.

Eighty Years Progress of the United States, Hartford, Conn.

Farmer and Planter, Pendleton, S. C., 1850-1860.

The Farmers' Register, Shellbanks and Petersburg, Va., 1833-1842.

Harper's New Monthly Magazine, New York, N. Y., 1850-1900.

Niles Weekly Register, Baltimore, Md., 1811-1849.

Once A Week, an Illustrated Miscellany of Literature, Art, Science and Popular Information, London, 1859-1880.

Soil of the South, Columbus, Ga., 1851-1857.

The Southern Cultivator, Augusta, Ga., 1843-1881.

VI. Statistical Collections:

Affleck's Southern Rural Almanac, Washington, Miss., 1854.

Hodgson, Joseph, *The Alabama Manual and Statistical Register,* Montgomery, 1869.

DeBow, J. D. B., *The Industrial Resources of the Southern and Western States,* New Orleans, 1852. 3 vols.

Donnell, E. J., *Chronological and Statistical History of Cotton,* New York, 1872.

Harper, Roland M., *Economic Botany of Alabama, Parts* I and II of *Geological Survey of Alabama,* Monograph No. 8, University, Alabama, 1928.

Hilgard, Eugene W., compiler, Report on Cotton Production in the United States including the Agricultural and Geographical features of the Several Cotton States (Special Report for Alabama by Eugene Allen Smith, State Geologist), *Tenth Census of the United States,* Washington, 1884.

Seybert, Adam, ed., *Statistical Annals,* Philadelphia, 1818.

Snedecor, V. Gayle, *A Directory of Greene County,* 1855-56, Mobile, 1856.

Stine and Baker, *Atlas of American Agriculture,* Washington, 1918.

Tharin, W. C., *A Directory of Marengo County for 1860-1861,* Mobile, n. d.

Watkins, James L., *King Cotton — A Historical and Statistical Review, 1790-1908,* New York, 1908.

Watkins, James L., *Production and Price of Cotton for One Hundred Years,* U. S. Dept. of Agriculture Bulletin, Miscellaneous Series No. 9, Washington, 1895.

Warden, D. B., *Statistical, Political and Historical Account of the United States of North America,* Edinburgh, 1819. 3 vols.

United States, *Fourth Census,* Washington, 1821.

United States, *Fifth Census,* Washington, 1831.

United States, *Sixth Census,* Washington, 1841.

United States, *Seventh Census,* Washington, 1853.

United States, *Eighth Census,* Washington, 1864.

VII. MISCELLANEOUS SOURCES:

Baldwin, Joseph G., *Flush Times in Alabama and Mississippi,* New York, 1853.

Bassett, John Spencer, *The Southern Plantation Overseer as Revealed by his Letters,* Northampton, Mass., 1925.

Catterall, Helen T., *Judicial Cases Concerning American Slavery and the Negro,* Washington, 1927-1932. 4 vols.

The Journal of Andrew Ellicott (Late Commissioner on Behalf of the United States During Part of the Year 1796, The Years 1797, 1798, 1799, and Part of the Year 1800), Philadelphia, 1814.

Hamilton, J. G. de Roulhac, ed., *The Papers of Thomas Ruffin,* Raleigh, 1920. 4 vols.

History of Navigation on the Tennessee River System, 75th Cong., 1st Sess., House Doc. 254, Washington, 1937.

Meek, A. B., *Romantic Passages in Southwestern History (including) Orations, Sketches, and Essays,* Mobile, 1857.

Mohr, Charles, *Plant Life of Alabama,* Montgomery, 1901.

Monette, John W., *History of the Discovery and Settlement of the Valley of the Mississippi,* New York, 1846. 2 vols.

Phillips, Ulrich B., ed., *Plantation and Frontier Documents, 1649-1863,* Cleveland, 1909. 2 vols. (First published as vols. I and II of *Documentary History of American Industrial Society,* Cleveland, 1909-1911. 11 vols.)

Turner, J. A., *The Cotton Planter's Manual: Being a Compilation of Facts from the Best Authorities on the Culture of Cotton,* New York, 1857.

Secondary Authorities

I. State and Local Histories:

Abernethy, Thomas Perkins, *The Formative Period in Alabama, 1815-1828,* Montgomery, 1922.

Alabama Historical Society Publications (Transactions), Tuscaloosa and Montgomery, 1855, 1897-1904. 6 vols.

Alexander, Claude H., *The Bank of the State of Alabama and its Branches,* unpublished M. A. Thesis, University of Alabama, 1928.

Avant, Gladys Beckham, *History of Washington County to 1860,* unpublished M. A. Thesis, University of Alabama, 1929.

Ball, Rev. T. H., *Clarke County — A Glance into the Great South-East from 1540-1877,* Grove Hill, Alabama, 1882.

Besson, J. A. B., *History of Eufaula, Alabama, the Bluff City of the Chattahoochee,* Atlanta, 1875.

Betts, Edward Chambers, *Early History of Huntsville, Alabama,* Montgomery, 1916.

Brewer, W., *Alabama: Her History, Resources, War Record, and Public Men,* Montgomery, 1872.

Brown, William Garrott, *A History of Alabama,* New York and New Orleans, 1905.

Comings, L. L. Newcomb and Albers, Martha M., *A Brief History of Baldwin County,* Fairhope, Alabama, 1928.

Craighead, Erwin, *Mobile: Fact and Tradition,* Mobile, 1930.

Dorman, Lewy, *Party Politics in Alabama from 1850 through 1860,* Wetumpka, Alabama, 1935.

Garrett, William, *Reminiscences of Public Men in Alabama for Thirty Years,* Atlanta, 1872.

Hamilton, Peter J., *Colonial Mobile,* Boston and New York, 1897.

Hardy, John, *Selma: Her Institutions and Her Men,* Selma, Alabama, 1879.

The Hill Country of Alabama, U. S. A., or the Land of Rest (Anonymous), London, 1878.

Jack, Theodore H., *Sectionalism and Party Politics in Alabama, 1819-1842,* Menasha, Wis., 1919.

Kennamer, John Robert, *History of Jackson County,* Winchester, Tenn., 1935.

Little, John Buckner, *History of Butler County, Alabama, from 1815-1855,* Cincinnati, 1885.

McCalley, Henry, *Northern Alabama, Its Topography, Geology, and Natural Resources,* n. p., n. d.

McCalley, Henry, *North Alabama or the Mountain, Manufacturing and Mineral Region of Alabama,* n. p., n. d.

McClellan, Captain R. A., *Early History of Limestone County,* Athens, Ala., 1881.

Memorial Records of Alabama, Madison, 1893. 2 vols.

Miller, L. D., *History of Alabama,* Birmingham, 1901.

Moore, Albert B., *History of Alabama and Her People,* Chicago and New York, 1927. 3 vols.

Neighbors, Rubye Grace, *Plantation Life in Hale County,* unpublished M. A. Thesis, University of Alabama, 1927.

Owen, Thomas McAdory, *History of Alabama and Dictionary of Alabama Biography,* Chicago, 1921. 4 vols.

Pickett, James Albert, *History of Alabama* (Randolph Edition) Sheffield, Alabama, 1896.

Riley, Rev. B. F., *History of Conecuh County, Alabama,* Columbus, Ga., 1881.

Robertson, W. G., *Recollections of Early Settlers of Montgomery County and their Families,* Montgomery, 1892.

Saunders, James Edmonds, *Early Settlers of Alabama with Notes and Genealogies by Elizabeth Saunders Blair Stubbs,* New Orleans, 1899.

Smartt, Eugenia Persons, *History of Eufaula, Alabama,* Birmingham, 1930.

Story, John N., *The History of Marshall County Prior to 1860,* unpublished M. A. Thesis, University of Alabama, 1930.

Smith, Nelson F., *History of Pickens County, Alabama, from its First Settlement,* Carrollton, Ala., 1856.

Street, O. D., *Marshall County One Hundred Years Ago,* Guntersville, Ala., 1903.

Teeple and Smith, publishers, *Jefferson County and Birmingham, Alabama, Historical and Biographical, 1887,* Birmingham, n. d.

Yerby, William Edward Wadsworth, *History of Greensboro, Alabama from its Earliest Settlement,* Montgomery, 1908.

II. Other Studies:

Abernethy, Thomas Perkins, *From Frontier to Plantation in Tennessee,* Chapel Hill, 1932.

The Alabama Historical Quarterly, Montgomery, 1930.

Alderman, Edwin A. and Gordon, Armistead C., *J. L. M. Curry, A Biography,* New York, 1911.

"Autobiography of Gideon Lincecum", *The Mississippi Historical Society Publications,* VIII.

Ballagh, James Curtis, "Southern Economic History: Tariff and Public Lands", *Annual Report of the American Historical Association, 1898,* Washington, 1899.

Bancroft, Frederic, *Slave-Trading in the Old South,* Baltimore, 1931.

Barnett, Gerald L., *Trade Relations Between Southern Illinois and the South, 1855 to 1870,* unpublished M. A. Thesis, University of Alabama, 1935.

Bassett, John Spencer, "The Industrial Decay of the Southern Planter," *The South Atlantic Quarterly,* II.

Beatty, Adam, *Southern Agriculture and the Best Method of Renovating the Soil,* New York, 1843.

Birney, William, *James G. Birney and His Times,* New York, 1890.

Bogart, Ernest Ludlow, *Economic History of American Agriculture,* New York, 1923.

Bogart, Ernest Ludlow, *The Economic History of the United States,* New York, 1912.

Boyd, Minnie Clare, *Alabama in the 'Fifties,* New York, 1931.

Brown, Harry Bates, *Cotton,* New York, 1927.

Cairnes, J. E., *The Slave Power: its Character, Career, and Probable Designs,* London, 1862.

Campbell, Richard L., *Historical Sketches of Colonial Florida,* Cleveland, 1892.

Campbell, T. J., *The Upper Tennessee*, Chattanooga, 1932.

Claiborne, J. F. H., "Trip through the Piney Woods", *The Mississippi Historical Society Publications*, IX.

Clark, T. D., "Live Stock Trade between Kentucky and the South, 1840-1860", *Register of the Kentucky State Historical Society*, XXVII, Frankfort, 1929.

Cleland, Herdman F., "The Black Belt of Alabama", *The Geographical Review*, X.

Cole, Dunham and Gras, eds., *Facts and Factors in Economic History*, Cambridge, Mass., 1932.

Collections of the Georgia Historical Society, Part 1, III, Savannah, 1848.

Collins, Winfield H., *The Domestic Slave Trade of the Southern States*, New York, 1904.

Cotterill, R. S., *The Old South*, Glendale, Calif., 1936.

Craven, Avery Odell, "Soil Exhaustion as a Factor in the Agricultural History of Virginia and Maryland", *University of Illinois Studies in Political Science*, XIII.

Dodd, William E., *The Cotton Kingdom*, New Haven, 1919.

Dewey, D. R., *Financial History of the United States*, New York, 1903.

Drewry, William Sidney, *The Southampton Insurrection*, Washington, 1900.

DuBose, John Witherspoon, *The Life and Times of William Lowndes Yancey*, Birmingham, 1892.

Flanders, Ralph Betts, *Plantation Slavery in Georgia*, Chapel Hill, 1933.

Foreman, Grant, *Indian Removal*, Norman, Okla., 1932.

Faulkner, Harold Underwood, *American Economic History*, New York, 1931.

Goodell, William, *The American Slave Code in Theory and Practice: Its Distinctive Features Shown by its Statutes, Judicial Decisions, and Illustrative Facts*, New York, 1853.

Gray, Lewis Cecil, *History of Agriculture in the Southern United States to 1860*, Washington, 1933. 2 vols.

The Gulf States Historical Magazine, Birmingham, 1903-1904. 2 vols.

Hall, A. R., *Early Erosion-Control Practices in Virginia*, United States Dept. of Agriculture, Miscellaneous Publication No. 256, Washington, 1937.

Hammond, M. B., *The Cotton Industry — An Essay in American Economic History*, New York, 1897.

Harper, Roland, "A Preliminary Soil Census of Alabama and West Florida", *Soil Science*, IV.

Harper, Roland, "Some Relations of Soil, Climate and Civilization in the Southern Red Hills of Alabama", *South Atlantic Quarterly*, XIX.

Hibbard, Benjamin Horace, *A History of Public Land Policies*, New York, 1924.

Kennedy, Renwick C., "Black Belt Aristocrats", Reprinted from *Social Forces*, Oct., 1934.

Knox, John Jay, *A History of Banking in the United States*, New York, 1911.

Loring, F. W. and Atkinson, C. F., *Cotton Culture and the South*, Boston, 1869.

Mallard, R. Q., *Plantation Life Before Emancipation*, Richmond, 1892.

Martin, William E., "Internal Improvements of Alabama", *Johns Hopkins University Studies in Historical and Political Science*, Series 20, No. 4.

McGrane, Reginald Charles, *The Panic of 1837*, Chicago, 1924.

McMaster, John B., *A History of the People of the United States, From the Revolution to the Civil War*, New York, 1910. 7 vols.

Paxson, Frederic L., *History of the American Frontier, 1763-1893*, New York and Boston, 1924.

Petrie, George, ed., *Alabama Polytechnic Institute Historical Studies*, Auburn, Ala. (various dates).

Phillips, Ulrich B., *American Negro Slavery,* New York, 1918.

Phillips, Ulrich B., *Life and Labor in the Old South,* Boston, 1929.

Phillips, Ulrich B., *History of Transportation in the Eastern Cotton Belt to 1860,* New York, 1908.

Phillips, Ulrich B., "The Economic Cost of Slave Holding", *Political Science Quarterly,* XX.

Phillips, Ulrich B., "The Origin and Growth of the Southern Black Belts", *American Historical Review,* XI.

Phillips, Ulrich B., "On the Economics of Slavery, 1815-1860", *Annual Report of the American Historical Association, 1912,* Washington, 1914.

Phillips, Ulrich B., "Conservatism and Progress in the Cotton Belt", Reprint from the *South Atlantic Quarterly,* January, 1904.

Phillips, Ulrich B., "The Economics of the Plantation", *South Atlantic Quarterly,* II.

Ramsdell, Charles W., "The Natural Limits of Slavery Expansion", *Mississippi Valley Historical Review,* XVI.

Rhodes, James Ford, *History of the United States from the Compromise of 1850 to the Final Restoration of Home Rule in the South in 1877,* New York, 1892-1919. 8 vols.

Roche, Emma Langdon, *Historic Sketches of the South,* New York, 1914.

Scherer, James A. B., *Cotton as a World Power,* New York, 1916.

Smedes, Susan Dabney, *Memorials of a Southern Planter,* Baltimore, 1888.

Smith, William A., *Lectures on the Philosophy and Practice of Slavery,* Nashville, 1856.

The South in the Building of the Nation, Richmond, 1909. 12 vols.

Stone, Alfred H., "The Cotton Factorage System of the Southern States", *American Historical Review,* XX.

Sydnor, Charles S., *Slavery in Mississippi,* New York and London, 1933.

Sydnor, Charles S., "Life Span of Slaves", *American Historical Review,* XXV.

Thacker, Page, *Plantation Reminiscences,* n. p., 1878.

Treat, Payson J., *The National Land System, 1785-1820,* New York, 1910.

Truett, Randle B., *Trade and Travel Around the Southern Appalachians Before 1830,* Chapel Hill, 1935.

Vance, Rupert B., *Human Geography of the South,* Chapel Hill, 1932.

Turner, Frederick J., "The South, 1820-1830", *American Historical Review,* XI.

Van Evrie, J. H., *Negroes and Negro Slavery. The First, An Inferior Race — The Latter, Its Normal Condition,* Baltimore, 1853.

Wilkinson, General James, *Memoirs of My Own Times,* Philadelphia, 1816. 3 vols.

Yarbrough, W. H., *Economic Aspects of Slavery in Relation to Southern and Southwestern Migration,* Nashville, 1932.

INDEX